Handbook of
Micromethods
for the
Biological Sciences

Handbook of Micromethods for the Biological Sciences

Georg Keleti, Ph.D.
and
William H. Lederer, Ph.D.

 VAN NOSTRAND REINHOLD COMPANY
New York Cincinnati Toronto London Melbourne

Van Nostrand Reinhold Company Regional Offices:
New York Cincinnati Chicago Millbrae Dallas

Van Nostrand Reinhold Company International Offices:
London Toronto Melbourne

Manufactured in the United States of America

Published by Van Nostrand Reinhold Company
450 West 33rd Street, New York, N.Y. 10001

Published simultaneously in Canada by Van Nostrand Reinhold Ltd.

15 14 13 12 11 10 9 8 7 6 5 4 3 2 1

Library of Congress Cataloging in Publication Data

Keleti, Georg.
  Handbook of micromethods for the biological
sciences.

  Includes bibliographical references.
  1. Biological chemistry—Technique. 2. Microbiol-
ogy—Technique. 3. Immunochemistry—Techniques.
I. Lederer, William H., joint author. II. Title.
QH345.K43   574.1'92'028   73-12027
ISBN 0-442-24290-5

To
the 60th birthday of Prof. Dr. Otto Westphal
G. K.

To
Joseph Lederer
and
Ferdinand W. Breth Foundation
W. L.

Preface

The procedures described in this book are methods from literature sources and are modifications utilized at the Max Planck Institut für Immunbiologie in Freiburg im Breisgau, Germany; Department of Biochemistry and Department of Microbiology, University of Pittsburgh School of Medicine, Pittsburgh, Pennsylvania; and Department of Biochemistry and Microbiology, University Comenius, Bratislava, Czechoslovakia.

This handbook has 106 procedures each specifically detailing a biochemical or microbiological method of preparation or analysis. They are written in a *concise, step-by-step* form including an objective and an evaluation, so that the researcher, technician or student can, without need of library search, analyze a biological preparation. Not only are all reagents specified, but the exact methods of preparation are delineated so that no further calculations are necessary. Illustrations are provided where required to clarify a method. Thus, the book is explicit in its methods and also advantageous because the analytical procedures are micromethods. However, they can easily be converted to macromethods.

The book is divided into three main chapters. The first is "Preparation of Material (primarily bacterial origin) for Chemical Analysis or Biological Characterization." The second chapter is "Chemical Microanalytical Methods," and the third chapter is "Biological Characterization."

Many of the procedures described in this book refer to lipopolysaccharide, but only as a model compound.

This handbook is written for microbiologists, biochemists and immunochemists. However, we specifically omitted procedures that are considered in the realm of molecular biology. It is addressed to universities, research institutes, industries and public health laboratories dealing with biological material. Its purpose is to be utilized as a laboratory bench guide for the researcher, technician or student.

The illustrations were drawn by Ilene Winn Lederer as modified from literature sources.

GEORG KELETI
WILLIAM H. LEDERER

vii

Contents

2. Microanalytical Methods

3. Biological Characterization

Handbook of
Micromethods
for the
Biological Sciences

1

Preparation of Material for Chemical Analysis or Biological Characterization

1 ACID HYDROLYSIS OF SUGAR DERIVATIVES USING DOWEX 50

Objective

This is a method that permits liberation of monosaccharides from sugar derivatives in preparation for further analysis of the free sugars.

Method

1. Add 5 mg of oligosaccharide into 500 μl of water.
2. Add activated Dowex 50 (procedure 3) to a concentration of 25% of the volume. *Shake.*
3. Seal the tube in a flame.
4. Place in a boiling water bath for 3 hours.

5. Cool to room temperature.
6. Centrifuge at 3000 rpm for 10 minutes at 4°C.
7. Dry the sample (supernatant) in a vacuum desiccator over $CaCl_2$ with NaOH in a beaker (procedure 12, Drying Samples in a Vacuum Desiccator).
8. Add 500 μl water.
9. Repeat steps 7 and 8.
10. Repeat step 7.

Reference

1. Keleti, J., H. Mayer, I. Fromme and O. Lüderitz, *Eur. J. Biochem.* **16**, 284 (1970).

2 ACTIVATION OF AMBERLITE IRA-410 RESIN

Objective

The activation and washing procedure is necessary to perform on commercially purchased Amberlite IRA-410 (HCO_3^- form) resin. The Amberlite resin is a useful and convenient neutralizing agent, e.g., see procedure 18, Hydrolysis of Lipopolysaccharides (H_2SO_4).

Method

1. Wash about 100 g of the resin in a Büchner funnel containing Whatman No. 1 filter paper with water until the pH of the resin is the same as the water.
2. Add 2 N HCl and allow to stand with occasional stirring for 10 to 15 minutes.
3. Wash with water until the pH of the resin is the same as the water.
4. Add 2 N NaOH and allow to stand with occasional stirring for 10 to 15 minutes.
5. Wash with water until pH of the resin is the same as the water.
6. Add saturated $NaHCO_3$ and allow to stand with occasional stirring for 20 minutes. Repeat this step three times.
7. Wash with water until pH of the resin is the same as the water.
8. Place resin on aluminum foil and cover with a weighing paper and allow to dry at room temperature for two to three days.
9. Store in a dark bottle for up to several months.

Reference

1. Personal communication and experience.

3 ACTIVATION OF DOWEX 50

Objective

This is a method for activation of Dowex 50 in preparation of using it for acid hydrolysis of oligosaccharides or glycosides (procedure 1) or reducing the pH from the alkaline region to between 4 and 5 after borohydride reduction in Preparation of Sugars for Gas-Liquid Chromatography (procedure 71).

Method

1. Use Dowex 50 W-X 200–400 Mesh, Hydrogen form (Bio-Rad Laboratories, 32nd and Griffin, Richmond, California).
2. Add 10-fold excess of 1 N HCl to Dowex 50 on a Büchner funnel (with Whatman No. 1 filter paper).
3. Let stand for about 15 minutes.
4. Apply the vacuum and wash the Dowex with water until the pH is that of the water.
5. Activate a supply to last for only one to two weeks.

Reference

1. Personal communication and experience.

4 BACTERIAL LIPOPOLYSACCHARIDES– GRAM-NEGATIVE (MODIFIED WESTPHAL)

Objective

The principle of this method is that bacteria after being suspended in a hot phenol-water mixture and then cooled, are separated into a water-soluble lipopolysaccharide (also polysaccharide) and nucleic acid layer and a phenol-soluble protein layer. It has recently been found that the phenol layer may also contain carbohydrates. The pellet after centrifugation contains cell debris.

4 ■ PREPARATION OF MATERIAL

Method

1. 20 g of dried bacteria (procedure 25, Preparation of Bacteria for Cell Component Isolation) are suspended in 350 ml water in a 68°C water bath.
2. 350 ml of 90% phenol (preheated to 68°C) is added, with vigorous stirring for 30 minutes.
3. Cool to about 10°C in an ice bath.
4. Centrifuge at 7000 rpm for 45 minutes.
5. Aspirate the upper water layer. (This contains the lipopolysaccharide, nucleic acids and may contain polysaccharides.)
6. To the phenol layer and insoluble residue is added another 350 ml of water. *Stir.* It is heated in a water bath at 68°C for 30 minutes and steps 3, 4 and 5 are repeated.
7. Combine the aqueous layers. (Use phenol layer for protein isolation, procedure 26, Protein Isolation.)
8. Dialyze the aqueous layers against water for three days with numerous changes (to remove the phenol).
9. Use a rotary evaporator (40°C under reduced pressure) to reduce volume to about 100 ml.
10. Centrifuge for 15 minutes at 5000 rpm and discard the pellet.
11. Lyophilize the supernatant. This is the crude lipopolysaccharide which may be further purified (see procedure 27, Purification of Lipopolysaccharide, Modified Westphal).

Reference

1. Westphal, O. and K. Jann, in *Methods in Carbohydrate Chemistry*, edited by R. L. Whistler, J. N. BeMiller and M. L. Wolfrom, vol. 5, p. 83, Academic Press, New York, 1965.

5 CAPSULAR POLYSACCHARIDE ANTIGEN PREPARATION

Objective

Acidic polysaccharides form a major constituent of the capsule around some bacterial strains. The polysaccharides can be released using differential ionic strength salts with quaternary ammonium salts such as Cetavlon (hexadecyltrimethylammonium bromide) and then isolated.

In the first salt-Cetavlon extraction the RNA salt is precipitated. The ionic strength of the supernatant is then decreased with water and after centrifugation a sediment containing the crude acidic polysaccharide-Cetavlon salt is formed. The polysaccharide is then converted to the sodium salt and isolated.

Method

1. 500 mg of lyophilized supernatant from step 2, procedure 27, Purification of Lipopolysaccharide.
2. Dissolve in 50 ml NaCl reagent A.
3. Add 25 ml reagent B.
4. Stir for 15 minutes at room temperature.
5. Centrifuge at 10,000 rpm for 30 minutes. (The pellet contains crude RNA Cetavlon salt.)
6. Add 225 ml water to the supernatant. *Mix.*
7. Let stand at 4°C overnight.
8. Centrifuge at 10,000 rpm for 30 minutes at 4°C. (The supernatant contains low molecular weight lipopolysaccharide.)
9. Dissolve the pellet in NaCl reagent C.
10. Add 10 volumes 95% ethanol.
11. Centrifuge at 10,000 rpm for 30 minutes.
12. The pellet contains the sodium salt of the capsular polysaccharide antigen.
13. Repeat steps 9, 10 and 11 on the pellet from step 12 to insure complete precipitation of the capsular polysaccharide antigen.
14. Dissolve the pellet in water.
15. Dialyze for two days at 4°C against water with several changes.
16. Lyophilize.

Reagents

A. 17.6 g of NaCl dissolved up to 1000 ml with water (0.3 M)
B. 4 g hexadecyltrimethylammonium bromide dissolved up to 100 ml with reagent A
C. 58.5 g of NaCl dissolved up to 1000 ml with water (1 M)

Reference

1. Westphal, O. and K. Jann, in *Methods in Carbohydrate Chemistry*, edited by R. L. Whistler, J. N. BeMiller and M. L. Wolfrom, vol. 5, p. 83, Academic Press, New York, 1965.

6 COMMON ANTIGEN

Objective

This is a method for preparing partially purified common antigen from several species of Enterobacteriaceae such as *E. coli*, *Enterobacter aerogenes*, *Salmonella*, and *Shigella*. These species contain in addition to the well-known somatic or 0 antigen, another antigen common to all. Ethanol in a concentration of 85% allows separation from crude culture supernatant fluids, the common antigen and the 0 antigen. The common antigen is ethanol-soluble and the 0 antigen is ethanol-insoluble. Common antigen antisera can be prepared by injecting rabbits with heat-killed suspensions of *E. coli* 014.

Method

1. Grow the Enterobacteriaceae strain for 18 hours at $37°C$ on brain veal agar (Difco) in Kolle flasks.
2. Suspend the growth of cells in phosphate hemagglutination buffer, pH 7.3 (Difco).
3. Heat the suspension in a boiling water bath for 1 hour.
4. Centrifuge at 23, 500 g for 20 minutes at $4°C$.
5. Suspend the supernatant in ethanol to a final concentration of 85% ethanol; the mixture is kept at room temperature for 18 hours.
6. Centrifuge at 23, 500 g for 20 minutes at $4°C$.
7. Evaporate the supernatant to a small volume in a rotary evaporator at $35°C$ (at reduced pressure).
8. Place the reduced volume in petri dishes and dry to a powder in a $37°C$ incubator.
9. Dissolve the powder in hemagglutination buffer, pH 7.3 (Difco) to the original volume of step 2.

Reference

1. Suzuki, T., E. A. Gorzynski and E. Neter, *J. Bacteriol.* 88, 1240 (1964).

7 COMPLEXING OF LIPOPOLYSACCHARIDES, GLYCOLIPIDS OR LIPID A WITH BOVINE SERUM ALBUMIN

Objective

This is a method for complexing endotoxin-like material with a protein. It is generally used to enhance the biological activity of endotoxin-like preparations. In this way new antigens can be created and the influence of the carrier on the biological activity can be evaluated.

Method

1. 20 mg lipid A, etc., is suspended in 10 ml water.
2. Add 10 μl of triethylamine.
3. Prepare a solution of 20 mg of bovine serum albumin (Fraction V Powder Fatty Acid Free) in 10 ml water.
4. Add the 2 solutions together and dry in a rotary evaporator at 40°C.
5. Add 15 ml of water and repeat the drying in a rotary evaporator at 40°C.
6. Add 20 ml water and incubate in a shaking water bath at 50°C for 30 minutes.
7. Treat in an ultrasonicator for 15 minutes.
8. Lyophilize (except for lipid A which should have the volume reduced on a rotary evaporator at 40°C and finally dried in a vacuum desiccator. See procedure 12, Drying Samples in a Vacuum Desiccator.)

Reference

1. Westphal, O., J. Gmeiner, O. Lüderitz and A. Tanaka, *Int. Convoc. on Immunol.*, Buffalo, N.Y. (1968), p. 33, Karger, Basel, New York, 1969.

8 DEPROTEINIZATION OF BIOLOGICAL MATERIAL (SOMOGYI)

Objective

This is a method to remove protein from biological material such as blood or tissue extracts. It is useful in that it is suitable for micromethods; it introduces no salts into the filtrate which is important if one evaporates the filtrate to a small volume and it precipitates anticoagulants such as fluoride and oxalate.

Method

1. 100 μl sample, e.g., blood, serum, plasma, etc.
2. Add 1.9 ml water. *Shake.*
3. Add 1 ml $Ba(OH)_2$ reagent A. *Shake.*
4. Add 1 ml $ZnSO_4$ reagent B. *Shake.*
5. Filter on a Büchner funnel through Whatman No. 1 filter paper (or centrifuge at 2000 rpm for 30 minutes).

Reagents

A. 3.6 g $Ba(OH)_2 \cdot 8H_2O$ dissolved up to 200 ml with water (1.8%)
B. 4.0 g $ZnSO_4 \cdot 7H_2O$ dissolved up to 200 ml with water (2%)

The solutions should neutralize each other. To test dilute 10 ml $ZnSO_4$ solution B with 50 ml water. Add 2 drops phenolphthalein. Titrate with $Ba(OH)_2$ solution A to a faint pink color; 10 ± 0.05 ml should be required.

Reference

1. Somogyi, M., *J. Biol. Chem.* **160**, 69 (1945).

9 DEPROTEINIZATION OF ENDOTOXIN

Objective

Deproteinization by the method of Somogyi is a useful procedure for removal of protein from the endotoxin-like material prepared by the Foster-Ribi method (procedure 13, Endotoxin, Foster-Ribi).

Method

1. 10 mg crude endotoxin (from step 20, procedure 13, Endotoxin).
2. Suspend in 5 ml water.
3. Add 10 ml 0.3 N Ba(OH)$_2$. *Shake.*
4. Add 10 ml reagent A. *Shake.*
5. Filter on a Büchner funnel through Whatman No. 1 filter paper.
6. Dialyze against water at 4°C with several changes until the dialyzate gives a negative reaction, i.e., no precipitate with saturated BaCl$_2$.
7. Evaporate to a smaller volume on a rotary evaporator (35°C at reduced pressure).
8. Lyophilize.

Reagent

A. 50 g ZnSO$_4$ · 7H$_2$O dissolved up to 1000 ml with water (5%)

References

1. Somogyi, M., *J. Biol. Chem.* **160**, 69 (1945).
2. Badakhsh, F. F. and J. W. Foster, *Amer. J. Vet. Res.* **31**, 359 (1970).

10   DETERMINATION OF RNA AND DNA

Objective

This is a method for separation and quantitation of RNA and DNA from tissue. It is a simple very accurate method.

Method

1. Homogenize 100 mg of tissue in 5 ml of ice-cold water for 4 minutes.
2. Add 2.5 ml of cold HClO$_4$ reagent A and let stand in an ice bath for 10 minutes.
3. Centrifuge at 10,000 rpm for 15 minutes at 4°C and discard the supernatant.
4. Wash the precipitate twice with 5 ml of cold HClO$_4$ reagent B.

5. Decant carefully the $HClO_4$ and allow the tube to drain onto a filter paper for a few minutes.
6. Add 4 ml of KOH reagent C and incubate in a shaking water bath at 37°C for 1 hour. (Make certain all material is dissolved.)
7. Chill the tube in an ice bath for 15 minutes.
8. Add 5 ml of cold reagent A.
9. Centrifuge at 10,000 rpm for 15 minutes at 4°C. Keep the supernatant.
10. Wash the precipitate twice with 5 ml of cold reagent B.
11. Combine the 3 supernatants and add them to a 50 ml volumetric flask. Dilute to the 50 ml mark with water.
12. Determine the RNA concentration at 260 mμ with a blank of 0.1 N $HClO_4$.
13. The precipitate containing the DNA is dissolved in 5.0 ml of KOH reagent C and incubated in a 37°C shaking water bath overnight. Dilute to exactly 15 ml with water.
14. The DNA is estimated by the method described in procedure 49.

Reagents

A. 24.4 ml of 70% $HClO_4$ diluted up to 500 ml with water (0.6 N $HClO_4$)
B. 8.13 ml of 70% $HClO_4$ diluted up to 500 ml with water (0.2 N $HClO_4$)
C. 150 ml of 1 N KOH diluted up to 500 ml with water (0.3 N)

Reference

1. Halliburton, I. W. and R. Y. Thomson, *Cancer Res.* **25**, 1882 (1965).

11   DNA PREPARATION

Objective

DNA is isolated from dried Gram-negative bacteria by means of sodium dodecyl sulfate. The DNA is then purified by chloroform–octanol and isopropanol.

Method

1. 30 g dried bacteria (procedure 25).
2. Suspend in 500 ml EDTA reagent A and 500 ml saline (0.15 M NaCl) at pH 8.
3. Mix in a 20°C water bath overnight.
4. Add 13.11 g sodium dodecyl sulfate.
5. Continue the stirring for 5 hours in a 20°C water bath.
6. Add 95% ethanol until a flocculant precipitate is formed.
7. Centrifuge at 10,000 rpm for 30 minutes.
8. Wash the precipitate twice with 70% ethanol (i.e., centrifuge at 6000 rpm for 25 minutes, 4°C).
9. Dissolve the precipitate by vigorous stirring in 600 ml NaCl reagent B in a blender.
10. Centrifuge at 15,000 rpm for 1 hour, 4°C, and discard the pellet.
11. Repeat steps 6, 7, 8, 9 and 10 twice. (NOTE: In step 9 use 600 ml saline [0.15 M NaCl] rather than reagent B, and in step 10 centrifuge at 10,000 rpm, rather than 15,000 rpm.)
12. To the supernatants add 500 ml chloroform: octanol (8:1).
13. Place in a separatory funnel. *Shake* and discard the lower organic phase.
14. Repeat steps 12 and 13 two more times.
15. Precipitate the saline solution with isopropyl alcohol.
16. Centrifuge at 10,000 rpm for 60 minutes at 4°C.
17. Wash the precipitate twice with 70% isopropyl alcohol (i.e., centrifuge at 6000 rpm for 25 minutes at 4°C).
18. Dissolve the precipitate in saline (0.15 M NaCl).
19. Add 20 mg ribonuclease.
20. Place in a 37°C water bath for 1 hour. *Shake* intermittently.
21. Repeat steps 19 and 20.
22. Add an equal volume of 90% phenol in a 20°C water bath, stir vigorously for 20 minutes.
23. Cool in an ice bath.
24. Centrifuge at 7000 rpm for 45 minutes.
25. Aspirate the upper water layer. (This contains the DNA.)
26. Add an equal volume of saline to the phenol layer and stir vigorously in a 20°C water bath for 20 minutes.
27. Repeat 23, 24 and 25 and combine the aqueous extracts.
28. Dialyze the aqueous extract against water for three days with numerous changes (to remove the phenol).

29. Precipitate with 70% isopropyl alcohol.
30. Centrifuge at 15,000 rpm for 30 minutes and discard the supernatant.
31. Wash the precipitate twice with 70% isopropyl alcohol (i.e., centrifuge at 10,000 rpm for 30 minutes at 4°C).
32. Dissolve the precipitate in saline (0.15 M NaCl).
33. Centrifuge at 45,000 rpm for 2 hours and discard pellet.
34. Recentrifuge at 65,000 rpm for 2 hours and discard pellet.
35. The supernatant contains the purified DNA.

Reagents

A. 37.22 g disodium EDTA dissolved up to 1000 ml with water (0.1 M), pH 8
B. 87.8 g NaCl dissolved up to 1000 ml with water (1.5 M)

Reference

1. Herrmann, R., *Zbl. Bakt. I. Abt. Orig. I.* **208**, 152 (1968).

12   DRYING SAMPLES IN A VACUUM DESICCATOR

Objective

This is a method for drying isolated material to a stable form. It is described because certain precautions are required in drying samples in a vacuum desiccator.

Method

1. Place sample in a refrigerator for 30 minutes.
2. Centrifuge at 2000 rpm, 4°C, for 10 minutes to release oxygen from sample.
3. Cover sample tubes with a single layer of cheesecloth (gauze) and secure it with a rubber band around the tubes.
4. Place sample in desiccator over $CaCl_2$ and NaOH or P_2O_5, or only $CaCl_2$ depending upon what is called for in the procedure.
5. Attach vacuum line from a water faucet vacuum aspirator to the sealed desiccator and slowly turn on the water. Then open the seal to the desiccator.

6. Allow sample to remain on the water faucet vacuum aspirator for 2 hours.
7. Close the vacuum seal on the desiccator; remove the vacuum line from the desiccator; and then turn off the water on the water faucet vacuum aspirator.
8. Attach a vacuum oil pump to the desiccator, turn on the vacuum oil pump and after a few minutes open the desiccator seal. (This vacuum pump system should contain a trap between the pump and desiccator. This trap should be cooled by a dry ice alcohol [or acetone] bath to trap condensed moisture.)
9. Desiccator is attached to the vacuum pump for 2 to 3 hours.
10. Seal the desiccator, remove the vacuum line and then turn off the vacuum pump.
11. If necessary, allow desiccator to remain sealed overnight.
12. Place a piece of filter paper on the opening of the desiccator to slow the air flow and slowly open the desiccator seal.

Note:
If the sample contains less than 200 μl of fluid it may dry in the vacuum desiccator after being on the water pump for 2 to 3 hours (step 7) and the further steps are unnecessary.

Reference

1. Personal communication and experience.

13 ENDOTOXIN (FOSTER-RIBI)

Objective

This is a method introduced by Ribi and co-workers utilizing a cold aqueous ether extraction of the bacterial cells. This yields a crude product, which, after several purification steps gives an endotoxin-like product from Gram-negative bacteria.

Method

1. Centrifuge the cultivated bacteria at 6300 rpm for 1 hour at 0°C.
2. Wash with cold water at 0°C.
3. Centrifuge at 7000 rpm for 30 minutes at 4°C and discard the supernatant.

4. Resuspend the pellet in water (prechilled to $4°C$) to the equivalent of about 5×10^{11} cells/ml (Klett Summerson colorimeter, filter no. 66, scale reading 500).

5. Add 2 volumes of cold ether and shake vigorously by inversion in a separatory funnel six times for 10 seconds each. After each 10 seconds of shaking remove the stopper to release pressure in the separatory funnel and wait 10 seconds between shaking.

6. Let the suspension stand overnight at room temperature in the separatory funnel in a hood.

7. Remove the lower aqueous phase and discard the ether supernatant.

8. Remove the ether in the aqueous phase by nitrogen aeration.

9. Centrifuge the aqueous phase at 6000 rpm, $0°C$, for 1 hour and discard the pellet. (The supernatant contains the crude endotoxin.)

10. Repeat step 10.

11. Dialyze the supernatant for five days at $4°C$ against distilled water with several changes of water.

12. Add sodium chloride to a final concentration of $0.15 M$ at $4°C$.

13. Slowly add ethanol, with continuous stirring at $4°C$, to a final concentration of 68%.

14. Let stand overnight at $4°C$.

15. Centrifuge at 2000 g for 45 minutes at $4°C$.

16. Resuspend the pellet at $4°C$ in $0.15 M$ NaCl to the same final volume as in step 13.

17. Repeat steps 14 to 16.

18. Suspend the pellet in water and dialyze against water for two days at $4°C$ with several changes of the water.

19. Place supernatant in a rotary evaporator ($35°C$ and reduced pressure) and evaporate to a small volume. (This material can be lyophilized and then deproteinized by procedure 9, if desired.)

*20. Centrifuge twice at 40,000 rpm (105,000 g) for 4 hours at $0°C$ and lyophilize the supernatants.

*21. Resuspend the pellet in a small volume of water and lyophilize.

References

1. Badakhsh, F. F. and J. W. Foster, *Amer. J. Vet. Res.* **31**, 359 (1970).
2. Ribi, E., W. T. Haskins, M. Landy and K. C. Milner, *J. Exp. Med.* **114**, 647 (1961).

*Both supernatant and pellet contain endotoxic activity.

14 ENDOTOXIN (MODIFIED BOIVIN)

Objective

This is one of the early methods for preparing endotoxin. The major component extracted from Gram-negative bacteria is a complex antigen containing polysaccharide, lipid and protein. The method was originally proposed by Boivin, and Mesrobeanu in 1933 but has been more recently modified by Anna Marie Staub in order to obtain a purer preparation.

Method

1. Grow Gram-negative bacteria in tryptose-phosphate broth for 48 hours.
2. Centrifuge cells at 7000 rpm for 30 minutes at 4°C.
3. Wash twice with a (0.85%) saline solution (7000 rpm, 30 minutes, 4°C).
4. Weigh the wet cells and suspend them in five times their weight with water at 4°C (1 g – 1 ml).
5. Add an equal volume of trichloroacetic acid reagent A. *Shake.*
6. Cool in an ice bath for 3 hours.
7. Let warm to room temperature.
8. Centrifuge at 7000 rpm for 30 minutes at 9°C and discard the pellet.
9. Neutralize the supernatant to pH 6.5 with NaOH (initially with 1 M NaOH and final adjustment with 0.1 M NaOH).
10. Cool to –4°C.
11. Precipitate with 2 volumes of 95% ethanol at –4°C by pouring the cold solution into 95% ethanol cooled to –10 to –15°C.
12. Let stand at –4°C overnight.
13. Centrifuge at –4°C, 7000 rpm for 30 minutes.
14. Dissolve the pellet in 0.1 of the original volume with water and neutralize to pH 7 if necessary.
15. Dialyze for three days at 4°C against water with frequent changing.
16. Centrifuge at 27,000 g, 4°C for 30 minutes and discard the pellet of microbial debris.
17. The supernatant may be lyophilized if desired.

Reagent

A. 81.7 g trichloroacetic acid dissolved up to 1000 ml with water (0.5 N)

Reference

1. Staub, A. M., in *Methods in Carbohydrate Chemistry*, edited by R. L. Whistler, J. N. BeMiller and M. L. Wolfrom, vol. 5, p. 92, Academic Press, New York, 1965.

15 ENZYME TREATMENT OF BIOLOGICAL MATERIAL

Objective

Enzymatic treatment of biological material is useful in elucidating the type of compound(s) responsible for biological activity, e.g., protein, nucleic acid, lipid, etc.

Method

1. 5 mg of dry weight of sample is dissolved in 5 ml of 0.15 M phosphate buffer reagents of the appropriate pH. *Homogenize.*
2. Add to each sample 2.5 mg of the following enzymes. *Shake.*
   A. ribonuclease, pH 7.6
   B. deoxyribonuclease, pH 5.0
   C. pepsin, pH 2.5
   D. lipase, pH 7.0
   E. lysozyme, pH 7.2
   F. trypsin (Difco Laboratories, 1/250), pH 8.2
   G. pronase (from *Streptomyces griseus* protease), pH 8.6
   H. control containing samples at above pH's but without enzyme
3. Place in a 37°C water bath for 3 hours. *Shake intermittently.*
4. Remove 2 ml from each sample for assay.
5. Centrifuge the remainder for 20 minutes at 12,000 g and 0°C.
6. Remove the supernatant and homogenize the pellet in 3.0 ml of saline or water. (The supernatant and sediment and material from step 4 from the pepsin-treated samples should be neutralized with 0.1 N NaOH.)

7. If necessary, the samples may be dialyzed against the buffer needed for the particular assay of sample activity.

Reagents

A. 10.2 g KH_2PO_4 dissolved up to 500 ml with H_2O (0.15 M)
B. 10.65 g Na_2HPO_4 dissolved up to 500 ml with H_2O (0.15 M)
C. Mix A and B solutions to obtain the appropriate pH buffers.

Reference

1. Bobo, R. A. and J. W. Foster, *J. Gen. Microbiol.* **34**, 1 (1964).

16 GLYCOLIPIDS FROM ROUGH STRAINS OF GRAM-NEGATIVE BACTERIA

Objective

This is an improved method for the preparation of glycolipids from Rough Strains of Gram-negative bacteria (mainly Enterobacteriaceae). The yields are higher than those of procedure 4, Bacterial Lipopolysaccharides—Gram-negative (Modified Westphal).

Method

1. After cultivation of the bacteria, centrifuge at 7000 rpm for 45 minutes.
2. Wash the pellet with water at 7000 rpm for 45 minutes.
3. Wash the pellet successively with 95% ethanol, acetone and ether at 9000 rpm for 10 minutes.
4. Dry in a vacuum desiccator over $CaCl_2$ to a constant weight.
5. To 25 g dried bacteria, 100 ml extraction reagent A is added.
6. Homogenize the material in a nonexplosive blender for 2 minutes at 4°C.
7. Stir for 10 minutes on a magnetic stirring plate at 4°C.
8. Centrifuge at 5000 rpm for 15 minutes at 4°C.
9. The supernatant containing the glycolipids is filtered through Whatman No. 1 filter paper into a round-bottomed flask.
10. Repeat extraction steps 5, 6, 7, 8 and 9 twice to the pellet and combine the supernatants.

11. Evaporate petroleum ether and chloroform in a rotary vacuum evaporator at 35°C.
12. Transfer to glass centrifuge tubes and add water dropwise until the glycolipids begin precipitating.
13. Centrifuge at 3000 rpm for 10 minutes at 9°C.
14. Decant the entire supernatant carefully.
15. Wipe the inside of the tube dry with a filter paper.
16. Wash the entire precipitate three times with only 2.5 ml of 80% phenol reagent B and wipe the inside of tube with filter paper after each decantation of the supernatant (7000 rpm for 10 minutes at 9°C).
17. Wash the precipitate three times with ether (9000 rpm for 10 minutes at 4°C).
18. Dry the precipitate in a vacuum desiccator over $CaCl_2$.
19. The dried glycolipid is suspended in 25 ml water and placed into a 45°C water bath for 10 minutes.
20. Immediately place sample in a desiccator and very carefully turn on the vacuum. (This step is performed to remove the dissolved air.) After 2 hours remove the sample from the desiccator.
21. Shake sample in a vibrator for 5 minutes at room temperature.
22. Centrifuge at 40,000 rpm for 4 hours at 4°C.
23. Discard the supernatant and dissolve the pellet in 20 ml water.
24. Centrifuge at 3000 rpm for 10 minutes at 4°C and discard the pellet of undissolved material.
25. Lyophilize the supernatant.

Reagents

A. 100 ml 90% phenol
   250 ml chloroform
   400 ml petroleum ether
B. 80% phenol is prepared by adding 89 ml of 90% phenol to 11 ml water.

Reference

1. Galanos, C., O. Lüderitz and O. Westphal, *Eur. J. Biochem.* **9**, 245 (1969).

17   HYDROLYSIS OF LIPOPOLYSACCHARIDES (HCl)

Objective

This is a method for releasing of amino sugars. The amino sugars can be further analyzed or isolated by paper electrophoresis, paper chromatography, or on an amino acid analyzer.

Method

1. 5 mg of lipopolysaccharide.
2. Dissolve in 250 μl water.
3. Add 250 μl 8 N HCl for stable amino sugars or 4 N HCl for labile amino sugars.
4. Seal the tubes by flame.
5. Place in a boiling water bath for 12 hours for stable amino sugars or for labile amino sugars, for 3 hours.
6. Centrifuge at 2000 rpm for 10 minutes at 4°C.
7. Aspirate carefully the supernatant.
8. Add 1200 μl water to the supernatant.
9. Cool to 4°C in a refrigerator for 1 hour.
10. Dry the sample carefully in a vacuum desiccator over $CaCl_2$ with NaOH pellets. (See procedure 12, Drying Samples in a Vacuum Desiccator.)
11. Add 150 μl water.
12. Dry again in a vacuum desiccator as described in step 10.

Note:

(a) stable amino sugars, e.g., glucosamine
(b) labile amino sugars, e.g., 3-amino sugars

Reference

1. Keleti, J., O. Lüderitz, D. Mlynarčík and J. Sedlák, *Eur. J. Biochem.* **20**, 237 (1971).

18 HYDROLYSIS OF LIPOPOLYSACCHARIDES (H_2SO_4)

Objective

This is a method to release pentoses, hexoses, deoxyhexoses, and uronic acids from lipopolysaccharide. The sugars may then be further isolated by procedure 19, Isolation of Monosaccharides, Oligosaccharides or Aminosugars from Paper Chromatograms.

Method

1. 5 mg lipopolysaccharide (see procedure 27, Modified Westphal).
2. Dissolve in 250 μl water.
3. Add 250 μl 2 N H_2SO_4.
4. Seal the tubes by flame.
5. Place in a boiling water bath for 4 hours.
6. Centrifuge at 2000 rpm for 10 minutes at 4°C.
7. Aspirate carefully the supernatant.
8. Add 1200 μl water to the supernatant.
9. Add activated Amberlite IRA–410 (HCO_3^- form, see procedure 2, Activation of Amberlite 410 Resin) until pH 7. *Shake*.
10. Allow the Amberlite to settle by gravity.
11. Aspirate the supernatant.
12. Cool to 4°C for 1 hour in a refrigerator.
13. Dry sample carefully in a vacuum desiccator over P_2O_5 (or $CaCl_2$), (see procedure 12, Drying Samples in a Vacuum Desiccator).

Reference

1. Keleti, J., O. Lüderitz, D. Mlynarčík and J. Sedlák, *Eur. J. Biochem.* **20**, 237 (1971).

19 ISOLATION OF MONOSACCHARIDES, OLIGOSACCHARIDES OR AMINO SUGARS FROM PAPER CHROMATOGRAMS

Objective

This is a method for isolation and purification of monosaccharides, oligosaccharides or amino sugars using preparative paper chromatography.

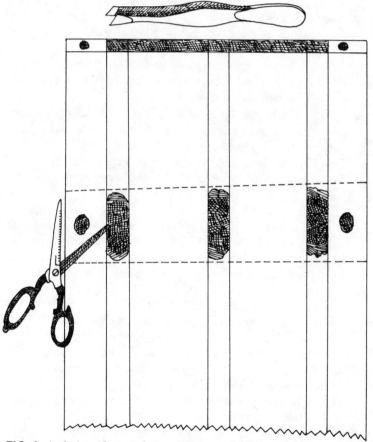

FIG. 1 Isolation of sugars from a Whatman 3 MM paper chromatogram.

Method

1. Run a paper chromatogram (using Whatman 3MM; see procedure 50, Estimation of Sugars by Descending Chromatography), adding about 50 to 100 μl hydrolyzed and neutralized sugar preparation at the origin of the paper chromatogram on a straight line. On the left side and right side of the paper at the origin, spot authentic standards (1 μl of 0.1 M sugars). (See Fig. 1.)
2. Draw with a pencil three 0.5 cm strips as shown in Fig. 1.
3. The following solvent systems may be used:
   A. neutral sugars
      n-butanol : pyridine : water (6 : 4 : 3 by volume)
   B. amino sugars
      ethyl acetate : pyridine : acetic acid : water (5 : 5 : 1 : 3 by volume) with ethyl acetate : pyridine : water (40 : 11 : 6 by volume) to saturate the atmosphere.
4. Dry the chromatograms in a hood for about 2 hours (until dry) and stain the three 0.5 cm strips, and the strip containing the spots of the standards as described in procedure 77, Stains of Thin Layer and Paper Chromatograms of Sugars by Ag$^+$/OH$^-$.
5. Cut the unstained portions of the unknown sugars using the three 0.5 cm strips as a guideline.
6. Use a trough containing water (or 0.01 N HCl for amino sugars) as an eluant. Place paper chromatogram between 2 microscope slides and allow the liquid to run by capillary action up the paper (see Fig. 2). Thirty seconds after the liquid reaches the top of the paper remove the paper and place it carefully with a tweezer on aluminum foil. Allow the top end of the paper to remain out of the aluminum foil (about 0.3 cm) and wrap the paper chromatogram like a cigarette in the aluminum foil. Place the so-called cigarette-wrapped paper chromatogram in a centrifuge tube (the top of the chromatogram which is sticking out of the aluminum foil 0.3 cm is placed about 2 cm from the bottom of the centrifuge tube [see Fig. 2]).
7. Tightly stopper the tube with a cork stopper so that the paper chromatogram cannot move.
8. Centrifuge at 2000 rpm, 4°C for 10 minutes.
9. Aspirate the liquid and repeat the centrifugation.

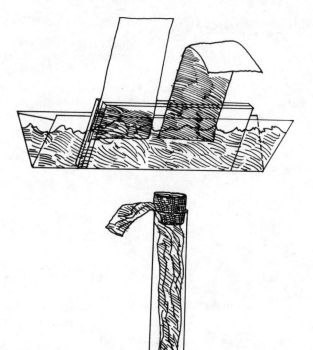

FIG. 2 Elution of the isolated sugars from strips of the paper chromatogram.

10. Pool the liquids and wash the centrifuge tube with 500 μl water and combine.
11. Dry the sample in a vacuum desiccator over $CaCl_2$.

Reference

1. Keleti, J., H. Mayer, I. Fromme and O. Lüderitz, *Eur. J. Biochem.* **16**, 284 (1970).

20 ISOLATION OF THE RIGID LAYER (MUREIN-SACCULUS)

Objective

The major constituent of the rigid inner layer of Gram-negative bacteria cell wall has been shown to be muramic acid derivatives. These constituents have been demonstrated to be responsible for the rigidity and shape of the bacterial cells. This is a method to isolate the rigid layer known as murein-sacculus from the bacteria.

Method

1. 30 g of frozen bacteria.
2. Suspend in 200 ml water.
3. Add 100 ml 0.1 N NaOH. *Stir*.
4. Bubble CO_2 into the solution until it is pH 7.
5. Add 50 mg DNAase. *Stir*.
6. Centrifuge at 10,000 g for 30 minutes. Discard the supernatant.
7. To the pellet, repeat steps 2, 3, 4, 5 and 6 twice more.
8. Suspend the pellet in 60 ml SDS reagent A.
9. Shake 10 ml portions in a Mickle-disintegrator with 0.17 mm diameter beads for 1 hour.
10. Suction off the liquid.
11. Centrifuge at 4000 g for 15 minutes, 4°C, and discard the pellet (small amounts of debris).
12. Centrifuge the supernatant at 22,000 g, 4°C, for 30 minutes.
13. Wash the sediment with water six times (22,000 g, 30 minutes, 4°C).
14. Add 50 ml water to the sediment. *Suspend well*.
15. As the suspension is stirring, add slowly and carefully 300 ml boiling SDS reagent B.
16. Let stand overnight at room temperature.
17. Centrifuge at 78,000 g for 45 minutes at 4°C.
18. Wash the pellet ten times with water (78,000 g for 30 minutes at 4°C).
19. Suspend the pellet in 50 ml phosphate buffer reagent C.
20. Add 5 mg pronase (i.e., 100 μg pronase/ml).
21. Stir in 60°C water bath overnight.

22. Centrifuge at 78,000 g for 30 minutes.
23. Wash the pellet eight times with water (78,000 g for 30 minutes at 4°C).
24. Dissolve pellet in a small volume of water.
25. Lyophilize.

Reagents

A. 4 g of sodium dodecyl sulfate (SDS) dissolved up to 1000 ml with water (0.4%)
B. 40 g of sodium dodecyl sulfate (SDS) dissolved up to 1000 ml with water (4%)
C. 1.36 g of KH_2PO_4 dissolved up to 100 ml with water and add 78.2 ml of 0.1 N NaOH. The pH should be 7.4.

References

1. Mardarowicz, C., Z. Naturforschg. 21b, 1006 (1966).
2. Martin, H. H. and H. Frank, Z. Naturforschg. 17b, 190 (1962).
3. Weidel, W., H. Frank and H. H. Martin, J. Gen. Microbiol. 22, 158 (1960).

21 LIPID A PREPARATION FROM LIPOPOLYSACCHARIDES OR GLYCOLIPIDS

Objective

Lipid A is a major constituent of endotoxin. It may be cleaved from lipopolysaccharides or glycolipids of Gram-negative bacteria by mild acid hydrolysis. Thus, water insoluble lipid A preparation may be obtained in relatively pure form.

Method

1. Suspend 100 mg of lipopolysaccharide or glycolipids in 6 ml water.
2. Add 6 ml of 0.2 N acetic acid. Mix vigorously.
3. Hydrolyze in a sealed tube in a boiling water bath for 4 hours.
4. Centrifuge for 25 minutes, 4°C at 13,000 g. (Use supernatant for

procedure 23, Partial Hydrolysis of Lipopolysaccharide and Related Material.)

5. *Wash* the pellet three times with water (13,000 g, 4°C for 25 minutes).
6. Wash the pellet once with acetone (13,000 g, 4°C for 25 minutes).
7. Dry the pellet in a desiccator above $CaCl_2$.

Reference

1. Galanos, C., E. Th. Rietschel, O. Lüderitz and O. Westphal, *Eur. J. Biochem.* **19**, 143 (1971).

22 MOLECULAR SIEVE CHROMATOGRAPHY (SEPHADEX OR SEPHAROSE)

Objective

This is a method to separate materials on the basis of their molecular weight. Materials with a greater molecular weight come out earlier from a column of Sephadex, lighter material later. The manner in which molecular weight is calculated is described in this procedure. The choice of the appropriate type of Sephadex depends on the molecular weight (and the chemical properties) of the substance. Each Sephadex type fractionates within a particular molecular weight range and molecules above this range are totally excluded from the gel. Molecules of a molecular weight in the range are usually eluted within an elution volume approximately equal to the total bed volume of the Sephadex.

Standards

A.

Protein	Molecular Weight	Wavelength
Trypsin inhibitor (from ox pancreas)	6,500	230 mμ
Ribonuclease	13,700	230 mμ
Chymotrypsinogen	25,000	230 mμ
Ovalbumin	45,000	230 mμ
Transferrin	90,000	230 mμ
Aldolase	150,000	230 mμ
Apoferritin	480,000	230 mμ
Thyroglobulin	669,000	230 mμ
α-crystallin	820,000	230 mμ

B. *Sieve type*	*Fractionation range–molecular weight (peptides and globular proteins)*
Sephadex	
G–10	up to 700
G–15	up to 1,500
G–25	1,000–5,000
G–50	1,500–30,000
G–75	3,000–70,000
G–100	4,000–150,000
G–150	5,000–400,000
G–200	5,000–800,000
Sepharose	
6B	10,000–3,000,000
4B	100,000–20,000,000
2B	10^6–10^8

Method

1. Allow Sephadex (or Sepharose) to swell in the buffer for the recommended time.
2. Pour column and equilibrate it with buffer.
3. Establish void volume with blue dextran (V_o) (5 mg/ml) (weight average–molecular weight 2×10^6; read at 625 mμ).
4. Apply sample.
5. Determine the volume at which your active fraction(s) elute from the column (V_e).
6. Apply at least 4 standards (10 mg/ml) to the column in runs of 2 standards per run to determine the elution volumes (V_e) of your standards.
7. Calculate K_{av} for your active fraction and the standards.

$$K_{av} = \frac{V_e - V_o}{V_t - V_o}$$

V_e = elution volume of the (active) material
V_o = elution volume of blue dextran
V_t = total volume of gel bed ($\cong \pi r^2 h$;
  r = radius of the column,
  h = height of the column)

8. Plot K_{av} (linear) of the standards versus molecular weight (logarithmically) on semi-logarithmic paper to prepare the standard curve.

9. From the K_{av} of your active material of your unknown, determine the molecular weight from the standard curve.

References

1. *Sephadex-gel filtration in theory and practice.* Pharmacia Fine Chemicals, Inc.
2. *Beaded Sepharose 2B–4B–6B.* Pharmacia Fine Chemicals, Inc., 800 Centennial Avenue, Piscataway, N.J. 08854
3. Andrews, P., *Methods of Biochemical Analysis*, **18**, 1, edited by D. Glick, Interscience Publisher, New York, 1970.

23 PARTIAL ACID HYDROLYSIS OF LIPOPOLYSACCHARIDES AND RELATED MATERIAL

Objective

This is a method to partially split lipopolysaccharides or other natural products by acid hydrolysis into smaller units such as oligosaccharides. These smaller units can then be isolated and further analyzed.

Method

1. Use supernatant from step 4, procedure 21, Lipid A Preparation from Lipopolysaccharides or Glycolipids. (The precipitated lipid A may be used for other procedures.)
2. Evaporate the supernatant with a rotary evaporator at 40°C under reduced pressure.
3. Dissolve in 1 ml water.
4. The degraded polysaccharide is precipitated by the addition of 10 ml of ethanol:acetone (1:1 by volume).
5. Centrifuge at 3000 rpm for 20 minutes at 4°C and discard the supernatant.
6. Wash the precipitate with ether (i.e., centrifuge for 10 minutes at 3000 rpm, 4°C).
7. Dry the precipitate in a vacuum desiccator over $CaCl_2$.
8. To each of 3 tubes is added 2 ml of 1 N H_2SO_4 and 20 mg of the

dried degraded polysaccharide (step 7) and the tubes are sealed by flame.

9. The tubes are placed in a boiling water bath for 7.5, 15 and 30 minutes respectively.
10. *Immediately* upon removal of the tubes from the boiling water the samples are neutralized with saturated $Ba(OH)_2$.
11. Centrifuge to remove the precipitated $BaSO_4$.
12. The oligosaccharides are separated by paper electrophoresis, procedure 80, Uronic Acid (Qualitative).
13. Stain the paper electropherograms as described in procedure 77, Staining of Thin Layer or Paper Chromatograms for Sugars by $Ag+/OH^-$.

Reference

1. Hämmerling, G., O. Lüderitz and O. Westphal, *Eur. J. Biochem.* **15**, 48 (1970).

24 PARTIAL ALKALINE HYDROLYSIS OF LIPOPOLYSACCHARIDES AND RELATED BIOLOGICAL MATERIAL

Objective

This is a method to partially split lipopolysaccharides or other natural products by alkaline hydrolysis into smaller units. These smaller units can then be isolated and further analyzed.

Method

1. To each of 5 tubes add 20 mg lipopolysaccharide and dissolve it in 2 ml of 0.1 N NaOH and seal the tubes by flame.
2. To a second set of 5 tubes add 20 mg lipopolysaccharide and dissolve it in 2 ml of 0.25 N NaOH and seal the tubes by flame.
3. Incubate 2 tubes from each set at 37°C and incubate the remaining 3 tubes from each set at 56°C.
4. The tubes incubated at 37°C are removed at 30 and 60 minutes. The tubes incubated at 56°C are removed at 15, 30 and 60 min-

utes. (This means that at each time point of both the 37°C and 56°C incubation, you remove 2 tubes, one hydrolyzed by 0.1 N NaOH and the other hydrolyzed by 0.25 N NaOH.)

5. Allow tubes to cool to room temperature.
6. Centrifuge at 3000 rpm for 10 minutes at 4°C.
7. Neutralize the supernatant with 1 N acetic acid.
8. Dialyze the neutralized supernatant against distilled water overnight.
9. Dry in a vacuum desiccator.

Reference

1. Beckmann, I., O. Lüderitz and O. Westphal, *Biochem. Z.* 339, 401 (1964).

25 PREPARATION OF BACTERIA FOR CELL COMPONENT ISOLATION

Objective

This is a method for harvesting, washing and drying bacteria to a stable form.

If one is not using pathogenic or dependent pathogenic bacteria, step 1 may be omitted.

Method

1. To the cultivated bacteria add 1.5% of 90% phenol and 1.5% of 95% ethanol and allow to remain at room temperature overnight.
2. Centrifuge for 45 minutes at 7000 rpm at 4°C.
3. Wash with water (7000 rpm, 30 minutes, 4°C).
4. Wash with 95% ethanol at 9000 rpm for 10 minutes at 4°C.
5. Wash with acetone at 9000 rpm for 10 minutes at 4°C.
6. Wash with ether at 9000 rpm for 10 minutes at 4°C.
7. Dry in a vacuum desiccator over $CaCl_2$ to a constant weight.

References

1. Keleti, J., D. Georch and D. Mlynarčík, *Präparative Pharmazie* 4, 170 (1968).

2. Galanos, C., O. Lüderitz and O. Westphal, *Eur. J. Biochem.* **9**, 245 (1969).

26 PROTEIN ISOLATION

Objective

This is a side product of the phenol–water method for extraction of lipopolysaccharide (see procedure 4, Bacterial Lipopolysaccharides). The major component of the phenol layer is protein which may be isolated by this method and further analyzed if desired.

Method

1. Use phenol layer from step 6, procedure 4.
2. Add 1000 ml 95% ethanol (if you began with 20 g of dried bacteria).
3. Add slowly, stirring constantly, sodium acetate reagent A until a sufficient precipitate has formed.
4. Centrifuge for 30 minutes at 7000 rpm.
5. Wash with an acetone:ether (1:1) solution (9000 rpm, 4°C, 10 minutes).
6. Wash with ether (9000 rpm, 4°C, 10 minutes).
7. Dry in a vacuum desiccator over $CaCl_2$.

Reagent

A. 100 g sodium acetate dissolved up to 1000 ml with water (10%)

Reference

1. Jann, B., Doctoral Dissertation, Freiburg i. Br., p. 34, 1965.

27 PURIFICATION OF LIPOPOLYSACCHARIDE (MODIFIED WESTPHAL)

Objective

Using this method one can purify the lipopolysaccharide from nucleic acids so that it is about 97% pure lipopolysaccharide. However, if further purification is required, the remaining 3% nucleic acids may be removed enzymatically (see procedure 15, Enzyme Treatment of Biological Material).

Method

1. Suspend the crude lipopolysaccharide (step 11, procedure 4, Bacterial Lipopolysaccharides) in water to give a 3% solution.
2. Centrifuge at 40,000 rpm (105,000 g) for 4 hours. (Lyophilize this first supernatant for further purification in procedure 5, Capsular Polysaccharide Antigen Preparation.)
3. Resuspend the pellet in water.
4. Repeat steps 2 and 3, two times.
5. Centrifuge for 10 minutes at 3000 rpm and discard the pellet.
6. Lyophilize the supernatant which contains the lipopolysaccharide and about 3% nucleic acids.

Reference

1. Westphal, O. and K. Jann, in *Methods in Carbohydrate Chemistry*, edited by R. L. Whistler, J. N. BeMiller and M. L. Wolfrom, vol. 5, p. 83, Academic Press, New York, 1965.

28   RNA PREPARATION

Objective

RNA may be isolated after the phenol–water extraction of procedure 4. RNA is found in the aqueous layer and is separated from the lipopolysaccharide and polysaccharide in relatively pure form.

Method

1. Use the reduced volume, centrifuged aqueous solution from step 9, procedure 4, modified Westphal.
2. Add ethanol–acetate reagent B until a flocculant precipitate is formed.
3. Centrifuge at 10,000 rpm for 30 minutes.

4. Wash pellet with 70% ethanol (i.e., centrifuge at 6000 rpm, 20 minutes, 4°C).
5. From the pellet prepare a 3% solution in water.
6. Centrifuge for 4 hours at 40,000 rpm (105,000 g) at 0°C and discard the pellet.
7. Repeat steps 2, 3 and 4.
8. From the pellet prepare a 1% solution in water.
9. Centrifuge for 6 hours at 50,000 rpm and discard any remaining pellet.
10. Place supernatant in an ice bath with a magnetic stirring plate below the ice bath.
11. Add 0.01 N HCl slowly (with constant stirring in an ice bath) until precipitation occurs.
12. Immediately after precipitation occurs, centrifuge at 6000 rpm for 30 minutes at 4°C.
13. Wash with 50% ethanol (i.e., centrifuge at 6000 rpm for 25 minutes at 4°C).
14. Dissolve the pellet in a small volume of water and lyophilize.

Reagents

A. 1. 84 ml 95% ethanol
   2. 14 ml water
   3. Add solid sodium acetate until a saturated solution is formed.
B. 1. 35 ml reagent A
   2. 1500 ml 95% ethanol

Reference

1. Herrmann, R., *Zbl. Bakt. I. Abt. Orig. I* 208, 152 (1968).

29 RNA PREPARATION (CETAVLON)

Objective

As a side product in the preparation of Capsular Polysaccharide Antigen (procedure 5), RNA may be isolated as the sodium salt. The principle of this method is described in procedure 5.

Method

1. Dissolve the pellet from step 5, procedure 5, Capsular Polysaccharide Antigen Preparation, in NaCl reagent A.
2. Add 10 volumes 95% ethanol.
3. Centrifuge at 10,000 rpm for 30 minutes at 4°C.
4. Repeat steps 1, 2 and 3 on the pellet from step 3.
5. Discard the supernatant.
6. Dissolve the pellet in water.
7. Dialyze for two days at 4°C against water with several changes.
8. Lyophilize.

Reagent

A. 58.5 g NaCl dissolved up to 1000 ml with water (1 M)

Reference

1. Westphal, O. and K. Jann, in *Methods in Carbohydrate Chemistry*, edited by R. L. Whistler, J. N. BeMiller and M. L. Wolfrom, vol. 5, p. 83, Academic Press, New York, 1965.

30 SEPARATION OF RNA, DNA AND PROTEIN IN TISSUE

Objective

This is a method for separation and preparation for quantitative analysis of RNA, DNA and protein in tissues. It is based on the preferential solubility of nucleic acids in hot trichloroacetic acid.

Method

1. 1 g of tissue is diluted up to 10 ml with ice-cold sucrose reagent A (10% w/v).
2. Homogenize tissue at 4°C.
3. Add 20 ml of ice-cold trichloroacetic acid reagent B.
4. Let stand for 30 minutes at 4°C.

5. Wash with 10 ml ice-cold reagent B. Centrifuge at 10,000 rpm for 20 minutes at 4°C. Discard the supernatant.
6. Suspend the precipitate in 10 ml of ethanol at room temperature.
7. Let stand for 30 minutes at room temperature.
8. Centrifuge at 10,000 rpm for 15 minutes at 4°C and discard the supernatant.
9. Repeat steps 6, 7, and 8.
10. Add 10 ml of ethanol: ether reagent C and heat in a 60°C water bath for 30 minutes.
11. Centrifuge at 10,000 rpm for 15 minutes at 4°C and discard the supernatant.
12. Suspend the precipitate in 2 ml of 1 N KOH.
13. Incubate in a 37°C water bath for 20 hours.
14. Add 0.4 ml of HCl reagent D and 2 ml trichloroacetic acid reagent E.
15. Let stand for 30 minutes at 4°C.
16. Centrifuge at 10,000 rpm for 20 minutes at 4°C.
17. The supernatant contains the hydrolyzed RNA and can be estimated by the orcinol procedure (procedure 67, Pentoses [Orcinol]).
18. Suspend the precipitate in 10 ml of trichloroacetic acid reagent E.
19. Heat in a 90°C water bath with occasional stirring for 20 minutes.
20. Cool to 4°C and let stand for 30 minutes at 4°C.
21. Centrifuge at 10,000 rpm for 20 minutes at 4°C.
22. The supernatant contains the DNA and it can be estimated by the diphenylamine reaction (procedure 43, 2-deoxysugars [diphenylamine]).
23. The proteins in the precipitate are estimated by the Lowry method (procedure 73).

Reagents

A. 8.55 g of sucrose is diluted up to 100 ml with water (0.25 M).
B. 10 g trichloroacetic acid is diluted up to 100 ml with water (10%).
C. 90 ml ethanol + 30 ml ethyl ether.
D. 25 ml concentrated HCl added to 25 ml water (6 N).
E. 5 g trichloroacetic acid diluted up to 100 ml with water.

References

1. Schneider, W. C., in *Methods in Enzymology*, edited by S. P. Colowick and N. O. Kaplan, vol. 3, p. 680, Academic Press, New York, 1957.
2. Trakatellis, A. C. and A. E. Axelrod, *Biochem. J.* **95**, 344 (1965).

31 SUCROSE DENSITY GRADIENT CENTRIFUGATION

Objective

This is a method for isolation of subcellular components and macromolecules based on their partial specific volume. Heavier substances are found in the denser sucrose at the bottom of the centrifuge tube and a gradient of less dense components occurs in upper regions of the tube. The method involves the determination of the equilibrium distribution of macromolecular material in the density gradient.

Method

1. Pipette 0.9 ml of 50% sucrose into six, 10 ml centrifuge tubes. Carefully layer on 0.9 ml 45% sucrose, 0.9 ml 40% sucrose, etc., into the centrifuge tubes. After 0.9 ml of 5% sucrose has been layered into the tubes add 0.9 ml water.
2. Place the centrifuge tubes at 4°C overnight and by spontaneous diffusion a linear gradient 5 to 50% will be obtained. (A linear gradient can also be prepared using a commercially available gradient-maker.)
3. Treat a 20% suspension of lyophilized biological material for 15 minutes in an ultrasonicator.
4. Layer on 0.1 ml of the treated biological material in water on top of the density gradient in the centrifuge tube.
5. Place the tube in a swinging bucket rotor (being certain that the rotor is balanced) and centrifuge at 100,000 g for 2 hours at 4°C.
6. Collect drops from the bottom of the centrifuge tube and assay the fractions. (Fractions can be collected by puncturing the bottom of the tube with a needle or by using a commercially available

gradient fractionater, e.g., Instrumentation Specialities Co., Inc., Lincoln, Nebraska.)

Note:

A continuous gradient can also be produced with cesium chloride, Ficoll, glycerol, etc.

References

1. Meselson, M., F. W. Stahl and J. Vinograd, *Proc. Nat. Acad. Sci. Washington*, **43**, 581 (1957).
2. Nowotny, A. *Basic Exercises in Immunochemistry*, p. 41, Springer-Verlag, New York, 1969.

2
Microanalytical Methods

32 ACETALDEHYDE

Objective

This is generally used in conjunction with periodate oxidation to elucidate the structure of an unknown sugar. One mole of acetaldehyde is formed from the periodate oxidation of a terminal methyl group vicinal to a hydroxyl group in a sugar such as a methylpentose, e.g., fucose, rhamnose, etc. Of course, glucose, ribose or sorbitol do not produce acetaldehyde upon periodate oxidation. The method is based on the formation of a rose-violet complex of acetaldehyde with p-hydroxybiphenyl with an absorption maximum of 560 mμ.

Standard

2 $\mu g/\mu l$ fucose or rhamnose, linear 1 to 4 μl

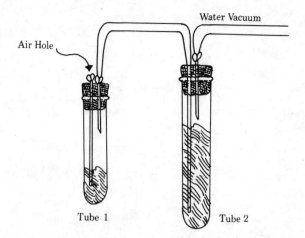

FIG. 1. Apparatus for acetaldehyde estimation. In tube 1 is the oxidation mixture (sample + NaIO₄). Tube 2 contains reagent B and after distillation also the acetaldehyde. The water vacuum regulates the amount of airstream taken into tube 1.

Method

1. 4 μl sample in tube 1 (see Fig. 1).
2. 50 μl reagent A (added to tube 1). *Shake.*
3. Into tube 2 add 1.19 ml reagent B.
4. Tube 1 is placed in a 70°C water bath. Slowly apply vacuum from a water pump aspirator to the outlet of tube 2. (Be certain that distillation and not merely aspiration is occurring.)
5. Remove tube 2 after 45 minutes and place in a boiling water bath for exactly 1.5 minutes. Cool to room temperature.
6. Read at 570 mμ.

Reagents

A. 1.07 g NaIO₄ dissolved up to 100 ml with water. Store in a dark bottle at 4°C (0.05 M).
B. 1. 150 μl water.
   2. 20 μl 12% CuSO₄ · 5H₂O. *Shake.* (1.2 g CuSO₄ · 5H₂O dissolved up to 10 ml with water)
   3. Add 1 ml cold concentrated H₂SO₄.

4. Place in an ice bath for 15 minutes.
5. 20 μl p-hydroxybiphenyl (150 mg dissolved up to 10 ml with 0.5 N NaOH).

References

1. Neidig, B. A. and W. C. Hess, *Anal. Chem.* **24**, 1627 (1952).
2. Kabat, E. A. and M. M. Mayer, *Experimental Immunochemistry*, 2d ed., p. 549, Charles C. Thomas, Springfield, 1961.

33 3-ACETYLAMINO SUGARS OR FREE 3-AMINO SUGARS*

Objective

3-Amino-3, 6-dideoxy-D-galactose (3-aminofucose) can be isolated from the lipopolysacchide of *Xanthomonas campestris*. 3-Amino-3, 6-dideoxygalactose and 3-amino-3, 6-dideoxyglucose (3-aminoquinovose) have been found in *E.coli*, *Salmonella*, *Arizona*, and *Citrobacter* strains. This is a method for quantitative estimation of 3-amino-3, 6-dideoxysugars and distinguishes them from 2-amino and 4-amino analogues.

Standard

2.5 μg/μl 3-acetylamino sugar, linear 1 to 10 μl

Method

1. Sample⁺.
2. 140 μl water.
3. 10 μl reagent A.
4. 20 μl reagent B. *Shake well.*
5. Centrifuge for 2 minutes at 2000 rpm. Incubate in a 55°C water bath for 20 minutes.
6. Immediately add 30 μl reagent C.

*The sugars are considered "free" after hydrolysis (e.g., see procedure 17).
⁺If sample is a free 3-amino sugar acetylation is required (see procedure 35, Acetylation of Free 3-Amino sugars) prior to analysis.

7. Let stand for 2 to 3 minutes at room temperature.
8. 160 μl reagent D. *Shake.*
9. Place in a boiling water bath for 3 minutes.
10. Cool in an ice bath.
11. 2.5 ml reagent F. *Shake.*
12. Incubate in a 37°C water bath for 20 minutes.
13. Read at 585 mμ.

Reagents

A. 6 ml of glacial acetic acid; add 10 ml water and adjust with 1 N NaOH to pH 4.5 and dilute up to 20 ml with water (5 M).
B. 912 mg $HIO_4 \cdot 2H_2O$ (periodic acid) dissolved up to 20 ml with water (0.2 M). Store in a dark bottle in the refrigerator.
C. 1.3 g $NaAsO_2$ dissolved up to 10 ml with water (1 M).
D. 2.26 g $K_2B_4O_7 \cdot 8H_2O$ (potassium borate) dissolved up to 20 ml with water (0.3 M).
E. 16 g p-dimethylaminobenzaldehyde dissolved up to 95 ml with glacial acetic acid + 5 ml concentrated HCl. This solution can be stored in a brown bottle at room temperature for one month.
F. Add 40 ml reagent E to 160 ml of glacial acetic acid (1:4). Prepare prior to use.

Reference

1. Ashwell, G., N. C. Brown and W. A. Volk, *Arch. Biochem. Biophys.* **112**, 648 (1965).

34 N-ACETYL AND O-ACETYL GROUPS

Objective

This is a simple and rapid colorimetric method for estimation of N-acetyl groups and indirectly O-acetyl groups. The principle of the method is a deacetylation with HCl methanol. The acetyl groups are converted to methyl acetate which is then estimated. With this method one cannot directly differentiate O-acetyl and N-acetyl groups. However, if a mild alkaline hydrolysis of the substance (30 minutes at 56°C in 0.25 N NaOH) is performed a saponification of the esters is achieved. Thus only the N-acetyl groups are now reactive in this method. Subtraction of the estimation before mild alkaline hydrolysis minus the N-acetyl estimation yields the O-acetyl group concentration.

Standard

A. 1. 2 μl ethyl acetate dissolved in 1 ml cold reagent A
   2. 10 to 100 μl of solution 1 dissolved up to 200 μl with cold reagent A (Start with step 11.)
B. 1 μg/μl N-acetylglucosamine ($\sim$0.005 M) dissolved in cold reagent A

Method

1. 0.1 to 1 μM sample (200 μl).
2. Evaporate to dryness in a vacuum desiccator over anhydrous CaCl$_2$ (using an oil pump to create a vacuum).
3. Add 50 μl reagent B.
4. Cool in a dry ice-alcohol bath for a few seconds, and rapidly seal the tube.
5. Place in a boiling water bath for 4 hours.
6. Cool to room temperature and open the seal of the tube.
7. Distill the sample in a microdistillation apparatus with a vacuum of 30 to 40 mm Hg, using dry ice alcohol to cool the collecting chamber (B) and heating the distillation tube (A) to 40°C in a water bath (Fig. 1).
8. After all the acid alcohol solution is distilled and collected in tube (B), repeat distillation with 50 μl absolute methanol added to the distillation tube (A).
9. After distillation remove collecting chamber (B) and place it in a 22°C water bath.
10. After 2 minutes, add 100 μl water.
11. Add 200 μl reagent C. *Shake.*
12. Let stand for 10 minutes at room temperature.
13. Add 200 μl reagent D. *Shake.*
14. Add 100 μl reagent F. *Shake.*
15. Read within 5 to 10 minutes at 520 mμ.

Note: When assaying a large number of samples it is convenient to keep the distillates in a freezer after stoppering and covering the sidearms with parafilm until they are to be analyzed.

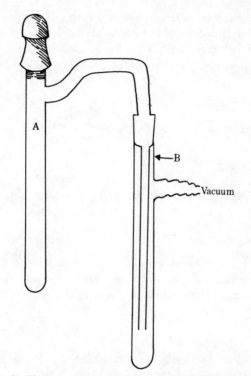

FIG. 1 Microdistillation apparatus for colorimetric determination of N-acetyl groups (A—distillation tube, B—suction chamber).

Reagents

A. 10 ml absolute methanol + 10 ml water (1:1).

B. 7.3 g dry HCl dissolved up to 100 ml with absolute methanol (2 N). Prepare fresh prior to use.

C. 1. 2.45 g hydroxylamine hydrochloride dissolved up to 100 ml with water (0.35 M). This solution is stored in a dark bottle in a refrigerator for up to one month.

   2. 6 g of NaOH dissolved up to 100 ml with water (1.5 M). Add 1 to 2 prior to use.

D. 6.5 ml of 70% perchloric acid dissolved up to 100 ml with water (0.75 M).

E. 3.5 ml 70% perchloric acid dissolved up to 100 ml with water (0.4 M).

F. Ferric perchloric acid solution is prepared by dissolving 1.9 g of $FeCl_3 \cdot 6H_2O$ in 5 ml of concentrated HCl. To this solution is added 5 ml of 70% perchloric acid. It is then evaporated in a boiling water bath almost to dryness. It is then diluted up to 100 ml with water.

Reference

1. Ludowieg, J. and A. Dorfman, *Biochim. Biophys. Acta* **38**, 212 (1960).

35 ACETYLATION OF FREE 3-AMINO SUGARS

Objective

Amino sugars are determined after acetylation with acetic anhydride. This is a method to acetylate free 3-amino sugars. The acetylated 3-amino sugars can be quantitatively determined.

Standard

1.5 µg/µl 3-amino-3,6-dideoxyglucose (or other available 3-amino sugar-0.01 *M*), linear 1 to 10 µl

Method

1. 40 µl sample from HCl hydrolysis (procedure 17, Hydrolysis of Lipopolysaccharides [HCl]).
2. 10 µl saturated $NaHCO_3$.
3. 10 µl reagent A. *Shake*.
4. Allow to stand for 10 minutes at room temperature.
5. Place in a boiling water bath for 3 minutes.
6. Cool to room temperature and if you want to quantitate continue with procedure 33, 3-Acetylamino sugars or Free 3-Amino sugars.

Reagent

A. 5 g acetic anhydride dissolved up to 100 ml with cold water in an ice bath (5%). Prepare prior to use.

Reference

1. Strominger, J. L., J. T. Park and R. E. Thompson, *J. Biol. Chem.*
   234, 3263 (1959).

36 N-ACETYLHEXOSAMINE

Objective

This is a Reissig, Strominger and Leloir modification of the method of
Aminoff, Morgan, and Watkins for the estimation of N-acetylamino
sugars. It is useful for analysis of easily hydrolyzable N-acetylamino
sugars in crude extracts of animal and bacterial tissues. It is also useful
for enzymatic studies and measurement of effluents from anion ex-
change column chromatography.

Standard

2.5 $\mu g/\mu l$ N-acetylglucosamine, or 5.0 $\mu g/\mu l$ N-acetylgalactosamine,
linear 1 to 10 μl

Method

1. 110 μl of water is added to the sample. (Prior to analysis the
   sample should be dried in a vacuum desiccator containing pellets
   of NaOH in a beaker over $CaCl_2$)
2. Add 110 μl $K_2B_4O_7 \cdot 8H_2O$ reagent A. *Shake.*
3. Place in a boiling water bath for 3 minutes.
4. Cool to room temperature with tap water.
5. Add 1 ml of reagent C. *Shake immediately.*
6. Incubate in a 37°C water bath for 20 minutes.
7. Cool to room temperature with tap water.
8. Read at 585 mμ.

Reagents

A. 2 g $K_2B_4O_7 \cdot 8H_2O$ dissolved up to 80 ml with water. Adjust pH
   to 9.1 with 0.1 N KOH and dilute to a final volume of 100 ml with
   water (2%).

B. 1. 10 g p-dimethylaminobenzaldehyde diluted up to 90 ml with glacial acetic acid.

2. 10 ml concentrated HCl. Store in a dark bottle at 4°C for up to one month.

C. Mix 2.5 ml of B + 22.5 ml glacial acetic acid (1:10). Prepare prior to use.

References

1. Reissig, J. L., J. L. Strominger and L. F. Leloir, *J. Biol. Chem.* 217, 959 (1956).

2. Aminoff, D., W. T. J. Morgan and W. M. Watkins, *Biochem. J.* 51, 379 (1952).

37 4-AMINO-L-ARABINOSE

Objective

This is a method for isolation and identification of a recently discovered amino sugar by long term mild-acid hydrolysis followed by electrophoresis and staining. This sugar is present in all strains of *Salmonella* and many strains of *E.coli* but it is unclear whether it is a component of the lipid A or saccharide part of lipopolysaccharide.

Standard

0.1 M glucose, glucosamine and glucose-6-phosphate

Method

1. 3 mg lipopolysaccharide is incubated in a 37°C water bath with 300 μl of 0.5 N HCl for 16 hours in a sealed tube.

2. Centrifuge at 2000 rpm for 10 minutes at 4°C.

3. Apply 90 μl of the supernatant of the sample to the origin on a high voltage paper electrophoresis apparatus. (Apply small volumes and dry after each application.)

4. Electrophoresis is performed for 60 minutes at 150 ma; 3000 volts using Whatman No. 1 electrophoresis buffer-pyridine:acetic acid:water (10:4:86 by volume), pH 5.3.

5. Air dry the strips of paper.
6. Stain one strip with ninhydrin (see procedure 75, Staining of Amino Sugars and Amino Acids) for amino sugars and a second strip with Ag^+/OH^- (see procedure 77, Staining of Thin Layer or Paper Chromatograms of Sugars by Ag^+/OH^-).

Evaluation

4-Amino-L-arabinose is a positively-charged sugar which migrates faster than glucosamine on paper electropherograms (M_{GlcN} 1.14). Glucose remains at the origin and glucose-6-phosphate is negatively-charged. The amino sugar turns yellow-brown after the paper electropherogram is dry, and is deep black after staining with Ag^+/OH^-, or is yellow-brown after staining with ninhydrin.

Since authentic 4-amino-L-arabinose is difficult to obtain one can use as presumptive evidence the presence of a ninhydrin and Ag^+/OH^- positive spot running slightly faster than glucosamine. With this presumptive evidence one may obtain small quantities of authentic 4-amino-L-arabinose from Professor W. A. Volk, School of Medicine, Charlotteville, Virginia.

Reference

1. Volk, W. A., C. Galanos and O. Lüderitz, *F.E.B.S. Letters* 8, 161 (1970).

38 AMINO SUGARS, AMINO ACIDS, ETC. (NINHYDRIN)

Objective

This is a method for quantitative estimation of amino sugars and other amino compounds.

Standard

leucine 1 $\mu g/\mu l$, linear 2 to 20 μl

Method

1. 100 μl sample in buffer reagent C.
2. 100 μl reagent A. *Shake*. Close tubes with Teflon-lined caps.
3. Place in a boiling water bath for 15 minutes. (Cover to protect from light.)
4. Add 1.5 ml ethanol:water (1:1 by volume).
5. Cool to about 30°C with a fan or with tap water. *Shake well.*
6. Read at 570 mμ.

Reagents

A. 1. 2 g ninhydrin (triketohydrindene hydrate).
   2. 300 mg hydrindantin.
   3. 75 ml methyl Cellosolve (ethylene glycol monomethyl ether).
   4. 25 ml reagent B. Bubble nitrogen gas through the solution. *Shake gently* and stopper tightly.
   5. Store in a dark bottle at 4°C (up to two weeks in a dark refrigerator).
B. 1. 544 g sodium acetate · 3H$_2$O dissolved up to 800 ml with warm water. Cool to room temperature.
   2. Add 100 ml glacial acetic acid.
   3. The pH can be adjusted to 5.5 with 50% NaOH.
   4. Dilute with water up to 1 liter.
C. 10 ml reagent B + 30 ml water.

Reference

1. Moore, S. and W. H. Stein, *J. Biol. Chem.* 211, 907 (1954).

39   BOROHYDRIDE REDUCTION

Objective

The borohydride reduction is used to reduce oligosaccharide samples. The free aldehyde groups will be reduced to sugar alcohols. After hydrolysis (e.g., procedure 18, Hydrolysis of Lipopolysaccharides [H$_2$SO$_4$]) the monosaccharides can be analyzed.

Reduction is also used prior to gas chromatography of the alditol acetates of the sugars by procedure 46, Determination of Sugars by

Gas-Liquid Chromatography or paper chromatography (see procedure 50, Estimation of Sugars by Descending Paper Chromatography) of sugars so that their alcohol derivatives may be compared to known alcohol standards.

Borohydride reduction is also commonly utilized in the identification of reducing terminal carbohydrates in oligosaccharides.

Method

1. 10 μl sample (about 0.1 M oligosaccharide).
2. 100 μl water.
3. 6 μl reagent A. *Shake*.
4. Let stand overnight in the dark at room temperature.
5. Add 1 N acetic acid until the pH of the reaction mixture is about 5 ($\sim$40 μl). *Shake*.
6. Control (the sugar should not be reduced in the reagent blank) 100 μl water.

   6 μl reagent A. *Shake*.

   40 μl 1 N acetic acid. *Shake*.

   10 μl sugar under investigation. *Shake*.
7. The boron compound from steps 5 and 6 can be removed by adding three times 1 ml methanol and evaporating on a rotary evaporator at 45°C.
8. If desired, periodate oxidation can be performed, procedure 68.

Reagent

A. 760 mg $NaBH_4$ dissolved up to 10 ml in 0.01 N NaOH in an ice bath.

References

1. Cope, A. C. and T. Y. Shen, *J. Amer. Chem. Soc.* **78**, 5912 (1956).
2. Hamilton, J. K. and F. Smith, *J. Amer. Chem. Soc.* **78**, 5907 (1956).

40 CARBOHYDRATES (FERRICYANIDE)

Objective

This is a method for quantitative estimation of carbohydrates. It gives similar results with all reducing carbohydrates and is suitable for amino sugars, aldohexoses and aldopentoses whereas procedure 41, analysis of carbohydrates by the phenol-sulfuric acid method, is unsuitable. However, prior hydrolysis of the sample is necessary (step 1) if the pure sugar is not utilized.

Standard

glucose 2 μg/10 μl, linear 1 to 8 μl (Glucose is dried in a vacuum desiccator over P_2O_5.)

Method

1. Hydrolyze a sample of 1 mg in a sealed tube in 100 μl 1 N H_2SO_4 for 4 hours. (Do not hydrolyze the pure sugar.)
2. Neutralize with 1 N NaOH.
3. 200 μl sample.
4. 200 μl carbonate-cyanide reagent A.
5. 200 μl ferricyanide reagent B. *Shake*.
6. Place in a boiling water bath for 15 minutes.
7. Cool to room temperature.
8. Add 1.0 ml ferric-iron reagent C. *Shake*.
9. Let stand for 15 minutes at room temperature.
10. Read at 690 mμ.

Reagents

A. Dissolve 1.33 g Na_2CO_3 and 162.5 mg KCN up to 250 ml with water.
B. Dissolve 125 mg $K_3Fe(CN)_6$ up to 250 ml with water. Store in a dark bottle in the refrigerator.
C. 375 mg $FeNH_4(SO_4)_2$ and 250 mg sodium lauryl sulfate dissolved up to 250 ml with 0.05 N H_2SO_4.

Reference

1. Park, J. T. and M. J. Johnson, *J. Biol. Chem.* **181**, 149 (1949).

41 CARBOHYDRATES (PHENOL-SULFURIC ACID)

Objective

This is a method for quantitative estimation of simple sugars, oligosaccharides, polysaccharides and their derivatives with free or potentially free reducing groups. It is advantageous in that no prior hydrolysis is required. However, it is not applicable with amino sugars, sugar alcohols, and is not sufficiently sensitive with pentoses. The method is particularly used for the estimation of sugars separated by paper chromatography.

Standard

dextran 10 $\mu g/\mu l$, linear 1 to 5 μl

Method

1. 200 μl sample.
2. 200 μl phenol reagent A.
3. Rapidly and carefully add 1 ml concentrated H_2SO_4. *Shake* (carefully to avoid burning yourself with the acid).
4. Let stand at room temperature for 10 minutes.
5. Let stand in a 30°C water bath for 20 minutes.
6. Read at 490 mμ. (The color is stable at room temperature for 2 to 3 hours.)

Reagent

A. 5.5 ml liquid phenol (90%) added to 94.5 ml water (5% final concentration).

Reference

1. Dubois, M., K. A. Gilles, J. K. Hamilton, P. A. Rebers and F. Smith, *Anal. Chem.* 28, 350 (1956).

42   6-DEOXYHEXOSES

Objective

This is a method for quantitative determination of 6-deoxyhexoses. The most common are fucose and rhamnose and interestingly generally in the L-form. This procedure is useful for the detection and determination of 2 to 10 μg of 6-deoxysugars (methylpentoses) in the presence of a considerable excess of other sugars.

Standard

L-rhamnose 2.0 μg/μl (0.2%), linear 1 to 5 μl

Method

1. 200 μl sample.
2. Cool in an ice bath.
3. While the sample is in an ice bath add 900 μl H_2SO_4 reagent A. *Shake carefully*.
4. Warm to room temperature.
5. Place in a boiling water bath for 5 minutes.
6. Cool to room temperature under running tap water.
7. Add 20 μl cysteine reagent B. *Shake*.
8. Let stand for 3 hours in the dark at room temperature. A greenish-yellow color appears which is stable for 24 hours.
9. Measure the difference between 396 and 427 mμ.

Reagents

A. 4 ml water + 24 ml concentrated H_2SO_4.
B. 0.3 g cysteine per 10 ml water (3%) prepared prior to use.

References

1. Dische, Z. and L. B. Shettles, *J. Biol. Chem.* **175**, 595 (1948).
2. Dische, Z. and L. B. Shettles, *J. Biol. Chem.* **192**, 579 (1951).

43 2-DEOXYSUGARS (DIPHENYLAMINE)

Objective

This is a quantitative colorimetric method for determination of 2-deoxysugars using diphenylamine reagent.

Standard

0.01 M 2-deoxyglucose or 2-deoxyribose, linear, 1 to 20 μl

Method

1. 300 μl sample.
2. 600 μl reagent A.
3. Heat for 12 minutes in a boiling water bath.
4. Measure at 510 mμ for 2-deoxyhexoses or at 595 mμ for 2-deoxypentoses.

Reagent

A. 1. 1 g diphenylamine* is dissolved in 100 ml glacial acetic acid.
   2. Add 2.75 ml H_2SO_4 (concentrated).

Reference

1. Dische, Z., in *Nucleic Acids*, edited by E. Chargaff and J. N. Davison, vol. 1, p. 287, Academic Press, New York, 1955.

44 2-DEOXYSUGARS (WEBB)

Objective

Deoxyribonucleic acid is hydrolyzed with trichloroacetic acid and reacted quantitatively with *p*-nitrophenylhydrazine. The product is separated from interfering substances and the optical density determined in alkaline solution. This method can also be utilized for qualitative and quantitative estimation of all 2-deoxysugars.

*Diphenylamine should be crystallized twice from 70% ethanol or petroleum ether.

Standard

0.01 M 2-deoxyribose, linear 1 to 10 μl

0.1 M 2-deoxyglucose, linear 1 to 5 μl

Method

1. 250 μl sample.
2. 750 μl 5% trichloroacetic acid (5 g/100 ml water).
3. 50 μl paranitrophenylhydrazine reagent A. *Shake well.*
4. Heat in a boiling water bath for 20 minutes.
5. Cool to room temperature (in a cold water bath or under running water).
6. Add 1 ml of n-butyl acetate. *Shake vigorously for 3 minutes.*
7. Centrifuge at 3000 rpm for 10 minutes to separate the organic from the aqueous layers.
8. Aspirate the upper butyl acetate layer and discard.
9. Repeat steps 6, 7, and 8.
10. Use 600 μl of the aqueous solution.
11. Add 400 μl 1 N NaOH. *Shake.*
12. Read at 560 mμ.

Reagent

A. 50 mg of paranitrophenylhydrazine dissolved up to 10 ml with absolute ethanol (0.5%). Prepare fresh daily.

Reference

1. Webb, J. M. and H. B. Levy, *J. Biol. Chem.* 213, 107 (1955).

45 DETERMINATION OF HYDROXY-FATTY ACIDS BY GAS-LIQUID CHROMATOGRAPHY

Objective

This is a method for qualitative and quantitative estimation of hydroxy-fatty acids by gas-liquid chromatography. Since the free hydroxy-fatty acids are unsuitable for gas-liquid chromatography the trimethyl silyl derivatives are prepared and utilized.

Standard (Internal)

β-hydroxymyristic acid

Method

1. Proceed from step 12 or if your material is already methylated proceed from step 21, procedure 56.
2. Add 500 μl KOH-dried pyridine.*
3. Add 100 μl hexamethyldisilazane. *Shake.*
4. Add 50 μl trimethylchlorosilane. *Shake.*
5. Heat in a 60°C water bath for 3 minutes.
6. Cool to room temperature.
7. Inject 1 μl of the sample into the gas chromatograph (using, e.g., a 3% OV-17, 3% OV-1, etc., 100/120 mesh column).
8. Determinations are performed as described in procedure 46, Determination of Sugars by Gas-Liquid Chromatography.

Reference

1. Yamakawa, T. and N. Ueta, *Jap. J. Exp. Med.* **34**, 37 (1964).

46 DETERMINATION OF SUGARS BY GAS-LIQUID CHROMATOGRAPHY

Objective

This is a method for qualitative and quantitative determination of sugars by gas-liquid chromatography.

Method

1. Use a suitable column, e.g., 3% OV-17, 3% OV-1, etc., 100/120 mesh, Gas Chrom Q (Applied Science Laboratories, Inc., P.O. Box 440, State College, Pennsylvania 16801).

*Add KOH pellets to a freshly-opened bottle of pyridine and store with the KOH pellets.

2. Program the temperature from 160 to 280°C to increase at the rate of 5°C per minute with an appropriate sensitivity setting.
3. Mix xylitol acetate (1 μg/μl in chloroform) with the sample to be tested at a ratio of about 1:1. (Use of xylitol acetate as the internal standard assumes xylose is not one of the unknown sugars in your sample.)
4. Inject 1 μl of the sample containing the internal standard (step 3) into the gas chromatograph.
5. After the run, inject 1 μl of reference standards (which have been converted to alditol acetates, procedure 71, Gas Liquid Chromatography of Sugars).
6. After the run of the standards, retention times are measured for each peak and compared to xylitol acetate and expressed as a relative retention time.

$$\text{Relative retention time} = \frac{\text{standard or unknown retention time}}{\text{xylitol acetate retention time}}$$

7. Measure the areas under the peaks (Area = $\frac{1}{2}$ base $\times$ height); determine the sum of the areas and calculate the percentage of the total of each component peak.
8. To determine the approximate weight of each component, compare the area of the peak with a peak of a known weight of xylitol acetate.[*]
9. For confirmation, run your unknown with known alditol acetate to ascertain that the peaks correspond.

[*]A useful formula is

$$\mu\text{g of sugar}/\mu\text{l} = R_{st} \times R_u \times I$$

where:

$$R_{st} = \frac{\text{the peak area of the internal standard in the standard run}}{\text{the peak area of the respective sugar in the standard run}}$$

$$R_u = \frac{\text{the peak area of the respective sugar in the sample run}}{\text{the peak area of the internal standard in the sample run}}$$

I = μg/μl of internal standard (xylitol acetate) added to sample and standard mixture

References

1. Sawardeker, J. S., J. H. Sloneker and A. R. Jeanes, *Anal. Chem.* **12**, 1602 (1965).
2. Burchfield, H. P. and E. E. Storrs, *Biochemical Applications of Gas Chromatography*, Academic Press, New York, 1962.
3. Fiereck, E. A. and N. W. Tietz, *Clin. Chem.* **17**, 1024 (1971).

47 3,6-DIDEOXYHEXOSES

Objective

Of the 8 possible 3,6-dideoxyhexoses only 5 have been isolated in cell walls of organisms, e.g., abequose, colitose, tyvelose, ascarylose, paratose. This is a modified method of Waravdekar and Saslaw for 2-deoxysugars in which by heating the sample at 55°C for 25 minutes with periodic acid 3,6-dideoxysugars can be determined.

Standard

1.5 μg/10 μl, linear 1 to 20 μl

Method

1. 50 μl sample in 0.5 N H_2SO_4.
2. Place in a boiling water bath for 45 minutes. (This step is omitted if using a pure deoxyhexose but is required when using a lipopolysaccharide.)
3. 50 μl water.
4. 100 μl H_5IO_6 reagent A. *Shake*.
5. Heat for 25 minutes in a 55°C water bath.
6. 200 μl arsenite reagent B. *Shake*.
7. 800 μl thiobarbituric acid reagent C. *Shake*.
8. Place in a boiling water bath for 12 minutes.
9. Measure while still warm the difference between 532 and 560 mμ.

Reagents

A. 570 mg $HIO_4 \cdot 2H_2O$ (periodic acid) dissolved up to 100 ml with reagent D (0.025 M). Store in a dark bottle in the refrigerator.

B. 2 g $NaAsO_2$ dissolved up to 100 ml with 0.5 N HCl (2%).

C. 300 mg 2-thiobarbituric acid dissolved up to 100 ml with water (0.3%). Heat in an 85°C water bath to obtain a light yellow-colored solution. Prepare prior to use.

D. Add 12.5 ml of 1 N H_2SO_4 to 87.5 ml of water.

References

1. Waravdekar, V. S. and L. D. Saslaw, *J. Biol. Chem.* **234**, 1945 (1959).
2. Cynkin, M. A. and G. Ashwell, *Nature* **186**, 155 (1960).
3. Lüderitz, O., personal communication, Freiburg, Germany, 1969.

48 ESTER-BONDED FATTY ACIDS (HYDROXYLAMINE)

Objective

This is a method for quantitative determination of ester-bonded lipids. Lipids which contain ester groups react with alkaline hydroxylamine (hydroxylaminolysis) to form a hydroxamic acid. Hydroxamic acid reacts with acidic ferric perchlorate to form an iron-chelate complex which is magenta and can be measured colorimetrically. Free fatty acids do not react in this method. Since different esters react at different rates this method is unsuitable for complex lipids because they require several hours to react in this system.

Standard*

methyl palmitate 2 mg/ml in chloroform, linear 10 to 100 μl

---

*If only acetyl groups are to be determined then ethyl acetate should be used as a standard.

Method

1. The standard and samples which are in chloroform are dried in a vacuum desiccator over $CaCl_2$. 100 μl acetone is added to all the samples and standards and the acetone dried in a vacuum desiccator over $CaCl_2$ (to insure complete removal of the chloroform).
2. 300 μl reagent A. *Shake.*
3. Heat for 2 minutes in a 65°C water bath.
4. Let stand at room temperature for 5 minutes.
5. 750 μl reagent C. *Shake.*
6. Let stand for 30 minutes at room temperature.
7. Read at 530 mμ.

Reagents

A. 1. 2 g hydroxylamine hydrochloride dissolved with 2.5 ml water (warm if necessary). Dissolve up to 50 ml with cold absolute ethanol.
   2. 4 g NaOH dissolved in 2.5 ml water (warm if necessary). Dissolve up to 50 ml with cold absolute ethanol.
   3. Add 5 ml solution 1 to 5 ml of solution 2. Mix in a stoppered graduated cylinder. Centrifuge at 3000 rpm for 10 minutes at 4°C and use supernatant within 1 hour after preparation.
B. 2.5 g ferric perchloric acid* (should not be yellow) dissolved in 5.0 ml of 70% perchloric acid ($HClO_4$) and 5.0 ml water. Dilute up to 50 ml with cold absolute ethanol. Store at 4°C.
C. 1. 2 ml reagent B.
   2. 1.5 ml 70% perchloric acid ($HClO_4$).
   3. Dissolve up to 50 ml with cold absolute ethanol. Prepare prior to use.

Reference

1. Snyder, F. and N. Stephens, *Biochim. Biophys. Acta* **34**, 244 (1959).

---

*Ferric perchloric acid can be purchased from the G. Frederick Smith Chemical Co., Columbus, Ohio.

49 ESTIMATION OF DNA (CERRIOTTI)

Objective

This is a method for quantitative estimation of DNA after separation by procedure 10, (Determination of RNA and DNA).

Standard

DNA 1 μg/μl (in 0.1 N KOH), linear 1 to 10 μl

Method

1. 500 μl sample.
2. 500 μl indole reagent C. *Shake.*
3. Place in a boiling water bath for 10 minutes.
4. Cool to room temperature under running tap water.
5. Extract three times with 1 ml of amyl acetate which removes interfering color into the amyl acetate. (The phases are separated by centrifugation and the upper amyl acetate layer is discarded.)
6. Read the aqueous layer at 490 mμ.

Reagents

A. 300 mg indole diluted to 500 ml with water (0.06%). Store in a refrigerator in a brown bottle.
B. 101.7 ml of concentrated HCl diluted to 500 ml with water (2.5 N).
C. Dilute 10 ml of reagent A with 10 ml reagent B. *Shake.* Prepare fresh daily.

References

1. Cerriotti, G., *J. Biol. Chem.* **198**, 297 (1952).
2. Keck, K., *Arch. Biochem. Biophys.* **63**, 446 (1956).

50 ESTIMATION OF SUGARS BY DESCENDING PAPER CHROMATOGRAPHY

Objective

This is a method for separation and identification of hydrolyzed or partially hydrolyzed products of biological material. It is useful for determining the sugars, oligosaccharides, amino sugars, or amino acids that are present. R_f values, which represent the distance migrated of the spot divided by the distance of the solvent front from the origin, may be calculated. Generally, the R_f of sugars is stated as a ratio of the R_f of glucose and are therefore designated as R_{fGl}.

Standard

0.1 M sugar

Method

1. Saturate the tank with the solvent by placing solvent in the upper trough and in the bottom of the tank and placing a blank Whatman No. 1 paper in the tank. It requires approximately 24 hours to fully saturate the tank.
2. Sugar preparations (e.g., see procedures 17 or 18, Hydrolysis of Lipopolysaccharides).
3. Add 15 μl water.
4. Spot 1 μl at the origin of Whatman No. 1 paper.
5. Fan dry the spots with a cold hair dryer.
6. Place the Whatman No. 1 paper in descending paper chromatography tank for about 18 to 26 hours.
7. Dry the paper at room temperature in a hood for about 2 hours (until dry).
8. Rechromatograph the paper as in step 6.
9. Dry as in step 7.
10. Stain appropriately (e.g., see procedures 75, 76 or 77).

Reagents

Some appropriate solvent systems are:
A. 1-butanol:ethanol:water (40:11:19 v/v)
B. Water:saturated phenol:1% ammonia:hydrocyanic acid (1:1:1:1 v/v)
C. 1-butanol:pyridine:water (10:3:3 v/v)
D. Ethyl acetate:acetic acid:water (3:3:1 v/v)
E. Ethyl acetate:acetic acid:formic acid:water (9:1.5:0.5:2.0 v/v)

Reference

1. Hough, L. and J. K. N. Jones, in *Methods in Carbohydrate Chemistry*, edited by R. L. Whistler and M. L. Wolfrom, vol. 1, p. 21, Academic Press, New York, 1962.

51 FORMALDEHYDE

Objective

This is generally used in conjunction with periodate oxidation (procedure 68) to elucidate the structure of an unknown sugar monosaccharide, or polysaccharide. One mole of formaldehyde is formed from the periodate oxidation of a primary alcohol group. For example, glucose yields 1 mole of formaldehyde, sorbitol, 2 moles of formaldehyde and fucose, no moles of formaldehyde. The periodate remaining during the oxidation is destroyed with arsenite and the formaldehyde determined with chromotropic acid.

Standard

D-sorbitol 1.82 $\mu g/10\mu l$ (0.001 M), linear 5 to 20 μl

Method

1. 20 μl sample.
2. 4 μl periodate reagent A. *Shake.*
3. Let stand at 4°C for 4 hours in a dark box.
4. 5 μl 4 N HCl.
5. 20 μl arsenite reagent B.
6. Let stand at room temperature for 8 minutes.
7. Add water to make 100 μl (i.e., add 51 μl water).
8. 1.0 ml chromotropic reagent C.
9. Place in a boiling water bath for 30 minutes. Cool to room temperature.
10. Read at 570 mμ.

Reagents

A. 1.07 g $NaIO_4$ (sodium metaperiodate) is dissolved up to 100 ml with water and stored in a dark bottle at 4°C (prepared accurately, 0.05 M).

B. 1.04 g $NaAsO_2$ (sodium metaarsenite) dissolved up to 20 ml with water (0.4 M).

C. 1. 100 mg chromotropic acid (4,5-dihydroxy-2,7-naphthalene-disulfonic acid disodium salt) dissolved up to 10 ml with water and filtered.

2. 30 ml concentrated H_2SO_4 is added to 15 ml of water and cool. Add about 40 ml of solution 2 into all of solution 1 so that a final volume of 50 ml is obtained. Store in a dark bottle at 4°C for up to three weeks.

References

1. Speck, J. C., Jr. and A. A. Forist, *Anal. Chem.* 26, 1942 (1954).
2. Kabat, E. A. and M. M. Mayer, *Experimental Immunochemistry*, 2d ed., p. 548, Charles C. Thomas, Springfield, 1961.

52 FORMIC ACID

Objective

This method is generally used in conjunction with periodate oxidation (procedure 68) to elucidate the structure of an unknown sugar. One mole of formic acid is formed from the periodate oxidation from secondary alcohols and from aldehyde groups. For example glucose yields 5 moles of formic acid, fucose, 4 moles of formic acid and fucositol, only 3 moles of formic acid.

Standard

sodium formate 680 μg/ml (0.01 M), linear 1 to 50 μl

Method

1. 175 μl sample (about 5 to 10 μl from periodate oxidation, procedure 68, is sufficient).
2. 175 μl reagent A. Preincubate for 2 minutes at 37°C.
3. 3 μl enzyme reagent B.
4. Place in a 37°C water bath for 10 minutes.
5. 700 μl perchloric acid reagent C.
6. Let stand for 25 minutes at room temperature.
7. Read at 350 mμ.

Reagents

A. 1. 11 mg tetrahydrofolic acid.
    2. 2 ml 2-mercaptoethanol, 1 M, pH 7 (7.8 g 2-mercaptoethanol + 75 ml water; adjust pH to 7 with 2 N KOH and bring to 100 ml with water).
    3. 0.5 ml water.
    4. 1.0 ml triethanolamine 1 M, pH 8. (14.9 g triethanolamine + 75 ml water; adjust to pH 8 with 2 N HCl and bring to 100 ml with water.)
    5. 0.5 ml ATP, 0.1 M (adenosine – 5′ triphosphate, disodium), (660 mg ATP + 10 ml water + 2 ml 1 N KOH).
    6. 1.0 ml MgCl$_2$ · 6H$_2$O (406 mg MgCl$_2$ · 6H$_2$O dissolved up to 20 ml with water).
B. Tetrahydrofolic formylase, from lyophilized cells of *Clostridium cylindrosporum* (1 μl = 100 units).
C. 2% HClO$_4$ (perchloric acid)–3 ml perchloric acid 60% + 87 ml water.

References

1. Rabinowitz, J. C. and W. E. Pricer, Jr., *J. Biol. Chem.* **229**, 321 (1957).
2. Rammler, D. H. and J. C. Rabinowitz, *Anal. Biochem.* **4**, 116 (1962).

53 FREE FATTY ACIDS (LONG CHAIN)

Objective

This is a colorimetric method for the quantitative estimation of long chain free fatty acids. It gives similar results as the gravimetric method (procedure 62, Isolation of Free Fatty Acids).

Standard

palmitic acid 4 $\mu g/\mu l$, linear 1 to 5 μl

Method

1. 1.0 ml sample in chloroform from isolation of free fatty acids (procedure 62, step 6).
2. 500 μl of copper reagent A. *Shake for 2 minutes.*
3. Centrifuge at 2000 rpm for 10 minutes at 4°C.
4. Aspirate upper aqueous layer and discard.
5. Remove 900 μl of the chloroform layer for assay into a dry tube. (Care is taken not to transfer any aqueous phase.)
6. Add 150 μl reagent B. *Shake.*
7. Read at 440 mμ.

Reagents

*A. 1. 18 ml 1 N triethanolamine reagent C.
   2. 2 ml 1 N acetic acid.
   3. 20 ml of a 6.45% $Cu(NO_3)_2 \cdot 3H_2O$.
*B. 0.1 gm sodium diethyldithiocarbamate up to 100 ml with *redistilled* 2-butanol (0.1% wt/vol).
 C. 1 N triethanolamine is prepared by adding 13.25 ml of triethanolamine up to 100 ml with water.

Reference

1. Duncombe, W. G., *Biochem. J.* 88, 7 (1963).

54 FRUCTOSE

Objective

Quantitative estimation of fructose in biological material is achieved by this method.

Standard

fructose 2 mg/ml in a saturated benzoic acid solution linear 2 to 6 μl

*Reagents A and B can be stored for 7 days at 4°C in dark bottles.

Method

1. 200 μl deproteinized sample (see procedure 8, Deproteinization).
2. 200 μl reagent A. *Shake.*
3. 600 μl reagent B. *Shake.*
4. Let stand in a 80°C water bath for 8 minutes.
5. Cool to room temperature under running water. Read at 530 mμ.

Reagents

A. 50 mg resorcinol dissolved up to 50 ml with 95% ethanol. Store in a dark bottle at 4°C for not longer than two months.
B. 50 ml of concentrated HCl is added carefully to 10 ml of water.

Reference

1. Roe, J. H., *J. Biol. Chem.* **107**, 15 (1934).

55 D-GALACTOSE (GALACTOSE OXIDASE)

Objective

This is a specific enzymatic method for the determination of galactose in biological material. If the material has a high protein content it is necessary to deproteinize the sample by procedure 8, Deproteinization (Somogyi), prior to the analysis of galactose.

Standard

D-galactose, 2 μg/μl (0.2%) in water, linear 2 to 10 μl

Method

1. 200 μl sample.
2. 200 μl Galactostat reagent A. *Shake.*
3. Incubate in a 37°C water bath for 1 hour.
4. 600 μl reagent C. *Shake.* (This stops the reaction.)
5. Read at 425 mμ.

Reagents

A. 1. Dissolve chromogen in 0.5 ml methanol.
   2. Add 30 ml water.
B. 1. Dissolve Galactostat (Worthington) in about 5 ml water (enough to dissolve it).
   2. Add Galactostat to chromogen reagent A.
   3. Add 5.0 ml of 2% Triton x-100 (Rohm and Haas Company).
   4. Dilute to a total volume of 50 ml with water. Store in a dark bottle at 4°C.
C. 1. 1.877 g glycine dissolved up to 100 ml with water (0.25 N).
   2. 1.0 g NaOH dissolved up to 100 ml with water. Add (1) to (2) to obtain a pH 9.7 glycine buffer (0.25 N).

References

1. Fischer, W. and J. Zapf, *Hoppe-Seyler's Z. Physiol. Chem.* **337**, 186 (1964).
2. Fischer, W. and J. Zapf, *Hoppe-Seyler's Z. Physiol. Chem.* **339**, 54 (1964).

56 GAS CHROMATOGRAPHY OF FATTY ACIDS FROM LIPOPOLYSACCHARIDE

Objective

This is a method for qualitative and quantitative estimation of free fatty acids. It is advantageous in that long chain free fatty acids are well separated.

Standard (Internal)

n-heptadecanoic acid (Start with step 13, methylation.)

Method

1. 10 mg of lipopolysaccharide or lipid A.
2. Hydrolyze in a sealed tube with 1 ml of 4 N HCl in a boiling water bath for 5 hours.

3. Cool to room temperature.
4. Add 1 ml chloroform and place in a separatory funnel.
5. Shake and remove the lower chloroform layer.
6. Repeat steps 4 and 5 three times.
7. Discard the upper aqueous layer.
8. Add to the combined chloroform extract a few grams of water-free sodium sulfate to remove any water.
9. Filter through Whatman No. 1 to remove the sodium sulfate.
10. Evaporate the filtrate to dryness in a vacuum rotary evaporator at 35°C.
11. Wash the dry flask with 4.0 ml of ether.
12. Dry with nitrogen in a 37°C sand bath.
13. Add 0.5 ml BF_3 methanol (Applied Science Laboratories, Inc., P.O. Box 440, State College, Pa., BF_3 methanol ester kit, 14% w/v, Catalogue No. 18017).
14. Stopper the flask tightly and place in a boiling water bath for 2 minutes.
15. Cool to room temperature.
16. Add 0.5 ml petroleum ether and 0.3 ml water. *Shake vigorously for 1 minute.*
17. Remove the upper phase of petroleum ether.
18. Repeat step 16 and 17.
19. Discard the lower aqueous phases.
20. Dry the combined upper petroleum ether phases in a 60°C sand bath with nitrogen.
21. Seal the tube and store at −20°C.
22. Add about 300 μl to 400 μl ethyl acetate (or chloroform) prior to chromatographic analysis.
23. Use an e.g., 3% OV-1, 3% OV-17, etc., column and proceed with determination as described for chromatography of sugars. (See procedure 46, Determination of Sugars by Gas-Liquid Chromatography.)

Note:
A. 1. If the sample is triglycerides and other lipids, add to 10 mg material 3.5 ml of 0.5 *N* methanolic NaOH.
2. Heat in a boiling water bath for 15 minutes.
3. Dry in a rotary evaporator at 35°C. Continue from step 13.
B. This procedure is also applicable to free fatty acids by beginning with step 13.

Reference

1. Metcalfe, L. D., A. A. Schmitz and J. R. Pelka, *Anal. Chem.* **38**, 514 (1966).

57 D-GLUCOSE (GLUCOSE OXIDASE)

Objective

This is a specific enzymatic method for the determination of glucose in biological material. If the material has a high protein content such as serum, it is necessary to deproteinize the sample by procedure 8, Deproteinization (Somogyi), prior to the analysis of glucose.

Standard

D-glucose (dextrose) 2.0 μg/μl (0.2%) in water, linear 1 to 10 μl

Method

1. 50 μl sample*.
2. 1 ml Glucostat reagent A. *Shake.*
3. Let stand 10 minutes at room temperature.
4. Add 5 μl HCl reagent B. *Shake.*
5. Let stand for a minimum of 5 minutes at room temperature (color is stable for several hours).
6. Read at 425 mμ.

Reagents

A. Glucostat (Worthington Biochemical Corporation).
    1. Add 30 ml water to a 50 ml graduate cylinder.
    2. Dissolve the chromogen in about 10 ml water.
    3. Add to graduate cylinder.

---

*If required, the sample can be hydrolyzed as described in procedure 18, Hydrolysis of Lipopolysaccharides (H_2SO_4).

4. Dissolve the contents of the Glucostat vial in about 5 ml water and add it to the graduate cylinder.
5. Adjust the volume to 50 ml with water.

B. Mix 5 ml concentrated HCl with 10 ml water (4 N).

Reference

1. *Glucostat Technical Bulletin (4–68)*, Worthington Biochemical Corporation, Freehold, New Jersey, 07728.

58 HEPTOSES

Objective

This is a general method for the quantitative analysis of heptoses. Heptose is determined by the cysteine-H_2SO_4 reaction. L-Glycero-D-mannoheptose is believed to represent a constituent of the common core structure of many O antigens of Gram-negative bacteria and this method also detects such compounds.

Standard

2.0 $\mu g/\mu l$ L-glycero-D-mannoheptose, linear 1 to 10 μl

Method

(Prepare a duplicate of both standards and samples.)
1. 200 μl sample.
2. Place sample in an ice bath.
3. 900 μl H_2SO_4 reagent A. *Shake* at 4°C.
4. After 3 minutes place in a 20°C water bath for 3 minutes.
5. Place sample in a boiling water bath for 10 minutes.
6. Cool to room temperature.
7. To one set of samples add 20 μl cysteine reagent B. To the other set of samples add 20 μl water.
8. Store samples in a dark cabinet for 2 hours at room temperature.
9. Measure the difference between 505 and 545 mμ.
10. Subtract the readings of the samples with water from the samples containing cysteine reagent B.

Reagents

A. 1. 5 ml water.
   2. Add 30 ml concentrated H_2SO_4, *carefully*.
B. 600 mg L-cysteine · HCl dissolved up to 20 ml with water (3%).

Reference

1. Osborn, M. J., *Proc. Nat. Acad. Sci.* **50**, 499 (1963).

59 HEXOSAMINE

Objective

The sample may be obtained after HCl hydrolysis (procedure 17) and is dried in a vacuum desiccator with $CaCl_2$ over NaOH (procedure 12). Total amino sugars are then determined after acetylation with acetic anhydride as described in this procedure.

Standard

2.0 $\mu g/\mu l$, glucosamine · HCl, linear 1 to 5 μl.

Method

1. 10 μl saturated $NaHCO_3$.
2. 10 μl reagent A. *Shake*.
3. 40 μl sample.
4. Incubate for 10 minutes at room temperature.
5. Place in a boiling water bath for 3 minutes.
6. Cool to room temperature.
7. 50 μl reagent B. *Shake*.
8. Place in a boiling water bath for 7 minutes.
9. Cool to room temperature.
10. 500 μl glacial acetic acid.
11. 200 μl Ehrlich reagent C. *Shake*.
12. Incubate in a 38°C water bath for 20 minutes.
13. Read at 585 mμ.

Reagents

A. 0.5 g acetic anhydride dissolved up to 10 ml with cold water ($4°C$) in an ice bath. Prepare prior to use.

B. 1 g $K_2B_4O_7 \cdot 4H_2O$ dissolved up to 20 ml with water (5%).

C. 1. 16 g p-dimethylaminobenzaldehyde dissolved up to 95 ml with glacial acetic acid.

    2. Add 5 ml concentrated HCl.

Reference

1. Strominger, J. L., J. T. Park and R. E. Thompson, *J. Biol. Chem.* **234**, 3263 (1959).

60 HEXOSES

Objective

This is a quantitative method for the determination of hexoses, e.g., glucose, mannose, allose, altrose, talose, galactose, idose, gulose, fructose, etc.

Standard

hexose 2.0 $\mu g/\mu l$ (0.2%), linear 1 to 5 μl

Method

1. 200 μl sample.
2. Cool in an ice bath.
3. While the sample is in an ice bath add 900 μl H_2SO_4 reagent A. *Shake carefully*.
4. Warm to room temperature.
5. Place in a boiling water bath for exactly 3 minutes.
6. Cool to room temperature under running tap water.
7. Add 20 ml cysteine reagent B. *Shake*.
8. Let stand for 30 minutes in the dark at room temperature.
9. Measure the difference between 415 and 380 $m\mu$.

Reagents

A. 4 ml water + 24 ml concentrated H_2SO_4.
B. 0.3 g cysteine per 10 ml water (3%) prepared prior to use.

Reference

1. Dische, Z., L. B. Shettles and M. Osnos, *Arch. Biochem. Biophys.* **22**, 169 (1949).

61   HEXOSES (ANTHRONE)

Objective

One of the methods to estimate the content of hexoses in biological material is the anthrone reaction. The principle of the reaction is the condensation of anthronol, which reacts with the sulfuric acid created furfural derivative of the sugars present. Amino sugars and N-acetyl-amino sugars do not react with this reagent. Hexuronic acids react only slightly with this reagent.

Standard

D-galactose and D-mannose 100 μg/ml. (100 mg of each D-galactose and D-mannose are dissolved in 200 ml water. Working solution is prepared daily by mixing a 1 : 10 dilution of the stock solution.)

Method

1. 500 μl sample.
2. Place in ice bath for 45 minutes.
3. Add dropwise 1 ml reagent A; shake in ice bath. Stopper tubes.
4. Place for 8 minutes in a 92°C water bath.
5. Reimmerse the tubes in an ice bath to stop the reaction.
6. Read at 620 mμ.

Reagent

A. 200 mg anthrone dissolved up to 100 ml with concentrated H_2SO_4 (0.2%). Prepare at least 4 hours prior to use and make fresh each day.

Reference

1. Shields, R. and W. Burnett, *Anal. Chem.* **32**, 885 (1960).

62   ISOLATION OF FREE FATTY ACIDS

Objective

This is a method for isolation of free fatty acids after acid hydrolysis of biological material. The free fatty acids are then estimated gravimetrically.

Method

1. 10 mg biological sample.
2. 1 ml 4 N HCl.
3. Seal tube with a flame.
4. Heat for 5 hours in a boiling water bath. *Shake* intermittently.
5. Extract three times with 1 ml chloroform.
6. Remove bottom chloroform layer using a separatory funnel.
7. Dry the chloroform layer in a previously weighed conical flask using a rotary evaporator (at reduced pressure) at 35°C.
8. Dry the conical flask in a vacuum desiccator over P_2O_5 until a constant weight is obtained.

Reference

1. Personal communication and experience.

63   2-KETO-3-DEOXYOCTONATE (KDO)

Objective

2-Keto-3-deoxyoctonate (KDO) was isolated in 1963 by Heath and Galambor and has been found to be a necessary constituent of the core of all Enterobacteriaceae lipopolysaccharide. Even the Re mutant e.g.,

S. minnesota R-595, which is defective in all other sugar components contains KDO.

Standard

2.5 μg/μl (~0.01 M), linear 1 to 3 μl

Method

1. 50 μl sample.
2. 50 μl 0.5 N H_2SO_4. *Shake.*
3. Place in boiling water bath for 8 minutes (this step is omitted if using pure KDO but is required when utilizing the lipopoly-saccharide). Cool to room temperature.
4. 50 μl H_5IO_6 reagent A. *Shake.*
5. Let stand for 10 minutes at room temperature.
6. 200 μl arsenite reagent B. *Shake well.*
7. 800 μl thiobarbituric acid reagent C. *Shake.*
8. Place in boiling water bath for 10 minutes.
9. Cool to room temperature under running tap water.
10. Add 1.0 ml butanol reagent D. *Shake.*
11. Centrifuge at 2,000 rpm for 10 minutes at 4°C.
12. Aspirate about 800 μl of upper butanol layer.
13. Measure the difference between 552 and 508 mμ in the butanol layer.

Reagents

A. 2.28 g H_5IO_6 (*o*-paraperiodic acid) dissolved up to 100 ml with water (0.1 N). (Store in a dark bottle.)
B. 4 g of $NaAsO_2$ dissolved up to 100 ml with 0.5 N HCl (4.0%).
C. 600 mg of 2-thiobarbituric acid dissolved up to 100 ml in boiling water (0.6%) and cooled to room temperature. Prepare prior to use.
D. 5 ml concentrated HCl added to 95 ml *n*-butanol.

References

1. Waravdekar, V. S. and L. D. Saslaw, *J. Biol. Chem.* **234**, 1945 (1959).
2. Lüderitz, O., personal communication, Freiburg, Germany, 1969.

64   LIPIDS (TOTAL)

Objective

This is a method for extraction and estimation of total lipids in material of biological origin. The homogenized biological material with a mixture of chloroform, methanol and water separate into 2 layers. The chloroform layer contains almost all the lipids which are estimated gravimetrically.

Method

1. 10 g biological material.
2. Homogenize in a blender for 2 minutes in a mixture of 10 ml chloroform and 20 ml methanol.
3. Add 10 ml chloroform, homogenize for an additional 1 minute.
4. Add 10 ml water and homogenize for 1 minute.
5. Filter on a Büchner funnel (using vacuum) through Whatman No. 1 filter paper. (When the residue is almost dry, pressure is applied with the bottom of a beaker to insure maximum recovery of solvent.)
6. Add filtrate to a 50 ml graduated cylinder.
7. Wash the filter paper and blender with a total of 10 ml chloroform and refilter.
8. Add this chloroform filtrate to the graduate cylinder.
9. Rewash the filter with 5 ml chloroform.
10. Add the chloroform filtrate to the graduate cylinder.
11. After allowing a few minutes for the phases to separate, the volume of the chloroform layer is recorded and the upper methanol-water layer removed by aspiration. (A small volume of the chloroform layer is also removed to ensure complete removal of the upper layer.)
12. The volume of the chloroform layer is recorded again.
13. Quantitatively add the chloroform into a weighed flask and evaporate in a 40 to 50°C water bath with a stream of nitrogen.
14. Dry the residue over phosphoric anhydride in a vacuum desiccator.
15. Weigh the flask a second time.

16. After weighing, 5 ml of chloroform is added three times and carefully decanted to remove the lipids.
17. The flask is dried as in step 14.
18. Weigh the flask a third time.
19. Weight of lipids = (step 15 – step 13)–(step 18 – step 13). (Step 13 is the weight of the empty flask.)
20. Total lipids =

$$\frac{\text{weight of lipid}}{\text{(step 19)}} \times \frac{\text{total volume of chloroform (step 11)}}{\text{volume of chloroform evaporated (step 12)}}$$

Reagents

Analytical grades of methanol and chloroform.

Reference

1. Bligh, E. G. and W. J. Dyer, *Can. J. Biochem. & Physiol.* **37**, 911 (1959).

65 NEURAMINIC ACIDS

Objective

Many tissues contain derivatives of neuraminic acid in combination with carbohydrates, lipids and proteins. The wide distribution of neuraminomucoproteins suggests that these materials have a protective function in mammalian cells. Chemical analysis has revealed that all *E. coli* with a K 1 serotype contain neuraminic acid. However, strains with other K serotypes contain only small amounts of neuraminic acid. The distribution and biological significance of neuraminic acid in bacteria has not yet been completely elucidated. This is a method for the detection and estimation of neuraminic acid in biological material.

Standard

N-acetylneuraminic acid 2.5 $\mu g/\mu l$, linear 10 to 50 μl

Method

1. 350 μl sample.
2. 700 μl reagent A. *Shake*.
3. 350 μl Ehrlich reagent B.
4. Cover tubes with marbles.
5. Heat in a boiling water bath for 30 minutes.
6. Cool rapidly to room temperature in an ice bath.
7. Read at 530 mμ.

Reagents

A. 6 g $Al_2(SO_4)_3 \cdot 18H_2O$ dissolved up to 20 ml with water (30%).
B. 1 g p-dimethylaminobenzaldehyde dissolved up to 20 ml with 6 N HCl. (Store in a dark bottle in the refrigerator.)

Reference

1. Barry, G. T., V. Abbott and T. Tsai, *J. Gen. Microbiol.* **29**, 335 (1962).

66   NUCLEIC ACIDS

Objective

Nucleic acids have a strong absorption in the ultraviolet region due to the purine and pyrimidine moieties. The absorption maximum is 260 mμ. The usual standard is purified yeast RNA and this method gives an approximate content of nucleic acids.

Standard

RNA or DNA 10 μg/μl in 0.01N NaOH, linear 1 to 5 μl

Method

1. 10 μl sample.
2. 990 μl 0.01 N NaOH. *Shake*.
3. Read at 260 mμ.

Reagent

0.01 N NaOH (40 mg NaOH pellets diluted to 100 ml with water).

Reference

1. E. A. Kabat and M. M. Mayer, *Experimental Immunochemistry*, p. 708. 2d ed., Charles C. Thomas, Springfield, 1961.

67 PENTOSES (ORCINOL)

Objective

This is a method for the quantitative analysis of pentoses—arabinose, ribose, lyxose and xylose.

Standard

Ribose 2 μg/μl (0.2%) in water, linear 1 to 10 μl.

Method

1. 250 μl sample.
2. 750 μl orcinol reagent A. *Shake*.
3. Heat for 25 minutes in a boiling water bath with a marble on top of each tube.
4. Cool to room temperature in cold water.

*5. Read the difference of 665 mμ and 550 mμ.

Reagents

A. 1. 1 g orcinol (recrystallized from benzene)**
   2. 375 mg $FeCl_3 \cdot 6H_2O$
   3. Dilute to 25 ml with water.
   4. Cool in an ice bath to 4°C.
   5. Add 475 ml of HCl reagent B.
   6. Store in a dark bottle in a freezer for not longer than six weeks.
B. Add 500 ml of concentrated HCl to 100 ml of water.

*Note: Interference due to precipitation may be removed with 1.5 ml butanol:pyridine:water (4:1:1). The tubes are shaken, centrifuged and the upper phase is removed for assay.

Reference

1. A. H. Brown, *Arch. Biochem.* 11, 269 (1946).

68 PERIODATE OXIDATION

Objective

One of the important methods to determine the structure of sugars is periodate oxidation. The principle of the reaction is the splitting of bonds between two C-atoms of a single chain carrying unsubstituted hydroxyl groups (vicinal glycol). For each split bond 1 mole of periodate is reduced to iodate. The consumption of periodate in moles may be followed spectrophotometrically by using the absorption band of metaperiodate which has a maximum of 220 to 240 mμ. After quantitative analysis of the split products one can elucidate the structure of a monosaccharide or polysaccharide. For example, 1 mole of fucose consumes 4 moles of periodate and the products are 1 mole of acetaldehyde and 4 moles of formic acid.

**Orcinol is purified by dissolving it in boiling benzene, decoloring with charcoal, and crystallizing after adding hexane.

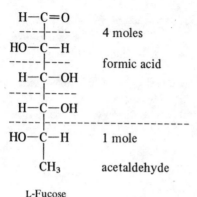

L-Fucose

If the aldehyde group of the sugar is reduced to a primary alcohol the consumption is again 4 moles of periodate and the products are 1 mole acetaldehyde, 3 moles of formic acid and 1 mole of formaldehyde.

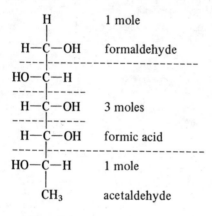

L-Fucositol

Standard

glucose 0.54 μg/μl (0.003 M)

Method

1. 100 μl water added to each of 4 tubes.
2. Withdraw 20 μl from tube 2 and tube 3.

3. Withdraw 40 μl from tube 4.
4. Add 20 μl of glucose or sample to tubes 2 and 4. *Shake.*
5. Add 20 μl periodate reagent A to tubes 3 and 4. *Shake.*
6. Let stand in the dark at room temperature for 0.5, 1, 2, 4, 6 and 8 hours for simple carbohydrates. If oligosaccharides or more complex polysaccharides are oxidized, 12, 24, 48 and 72 hour samples should be utilized in addition to the earlier time intervals.
7. Very accurately remove 10 μl from each tube at the various times and add it to 1.0 ml water. (This dilution stops the reaction.)
8. Read immediately at 225 mμ.

Evaluation

Use 4 tubes*, the second containing sugar but no periodate, the third containing periodate but no sugar and the fourth containing the sugar and periodate (as described in the procedure). Add after each time interval the optical densities of tubes 2 and 3 (blanks) and subtract tube 4 (reaction). Since glucose consumes 5 moles of periodate compare your unknown to glucose and calculate stoichiometrically the amount of periodate consumed.

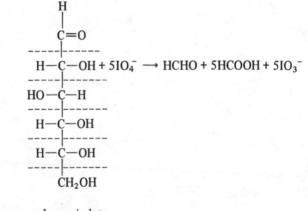

$$\text{Ratio} = \frac{\text{moles periodate}}{\text{moles glucose}}$$

*Tube 1 is used to blank the spectrophotometer to zero.

Reagent

A. Dissolve 535 mg $NaIO_4$ up to 100 ml with water (0.025 M).

References

1. Dixon, J. S. and D. Lipkin, *Anal. Chem.* 26, 1092 (1954).
2. Kabat, E. A. and M. M. Mayer, *Experimental Immunochemistry*, 2d ed., p. 542, Charles C. Thomas, Springfield, 1961.

69 PHOSPHORUS (INORGANIC)

Objective

This is a method for measuring inorganic phosphorus and labile phosphorus. The phosphorus reacts with ammonium molybdate creating phosphomolybdate which, after reduction with ascorbic acid, gives a blue-colored complex.

Standard

1.2 $\mu g/\mu l$ NaH_2PO_4 (molecular weight-120), linear 1 to 5 μl

Method

1. 100 μl sample.
2. 900 μl reagent A. Mix and cap with parafilm.
3. Heat for 2 hours in a 37°C water bath.
4. Read at 820 mμ.

Reagents

A. 1. 2 ml reagent B.
   2. 12 ml water.
   3. 2 ml 2.5% ammonium molybdate.
   4. 2 ml reagent C.
   Prepare reagent A prior to use.
B. 10 N H_2SO_4. (Add 200 ml H_2SO_4 [concentrated] to 520 ml water, *slowly and carefully*.)

C. 2 g ascorbic acid diluted up to 20 ml with water (10%). Prepare prior to use.

References

1. Chen, P. S., Jr., T. Y. Toribara and H. Warner, *Anal. Chem.* **28**, 1756 (1956).
2. Lowry, O. H., N. R. Roberts, K. Y. Leiner, M. L. Wu and A. L. Farr, *J. Biol. Chem.* **207**, 1 (1954).

70 PHOSPHORUS (TOTAL)

Objective

This is a method for measuring organically-bonded phosphorus after conversion into inorganic phosphorus. This inorganic phosphorus then reacts with ammonium molybdate creating phosphomolybdate which after reduction with ascorbic acid gives a blue-colored complex.

Standard

1.2 $\mu g/\mu l$ NaH_2PO_4 (molecular weight–120), linear 1 to 5 μl

Method

1. Dry 100 μl of the sample and standard in a vacuum desiccator containing $CaCl_2$.
2. After drying add 100 μl of ashing mixture reagent A.
3. Heat in an oven at 95°C for 2 hours.
4. Immediately heat in an oven at 165°C for 2 hours. Cool to room temperature.
5. Add 1 ml reagent D. *Mix immediately and thoroughly.*
6. Place in a water bath at 37°C for 2 hours.
7. Read at 820 mμ.

Reagents

A. 1. 20 ml water.
   2. 6.5 ml of 70% $HClO_4$.

3. 50 ml of 20 N H_2SO_4 (40 ml concentrated H_2SO_4 added to 32 ml of water).
4. Dilute to 100 ml with water.

B. 1. 2 ml sodium acetate $(NaC_2H_3O_2 \cdot 3H_2O) - 1$ M, (1.36 g $NaC_2H_3O_2 \cdot 3H_2O$ dissolved up to 10 ml with water).
2. 2 ml ammonium molybdate – 2.5% (500 mg ammonium molybdate dissolved up to 20 ml with water).
3. 16 ml water.

C. 2 g ascorbic acid dissolved up to 20 ml with water (10%). Prepare prior to use.

D. Add in an ice bath 9 ml reagent B + 1 ml reagent C. (After mixing B and C keep in an ice bath and use within 1 hour.)

Reference

1. Lowry, O. H., N. R. Roberts, K. Y. Leiner, M. L. Wu and A. L. Farr, *J. Biol. Chem.* **207**, 1 (1954).

71 PREPARATION OF SUGARS FOR GAS-LIQUID CHROMATOGRAPHY

Objective

This is a method to prepare from lipopolysaccharides or sugars their alditol acetate derivatives for gas-liquid chromatography.

Standard (Internal)

xylitol acetate (or as prepared below by borohydride reduction and acetylation of xylose.)

Method

1. If lipopolysaccharide is used sulfuric acid hydrolysis (see procedure 18, Hydrolysis of Lipopolysaccharides) is necessary, otherwise use 0.5 mg of known or unknown sugar.
2. Dissolve in 1 ml water.
3. Add 5 mg $NaBH_4$ and let stand for 3 hours in the dark.

4. Add fresh activated Dowex 50 H^+ (see procedure 3, Activation of Dowex 50). *Shake.* Attain a pH of 4 to 5.
5. Filter through Whatman No. 1 into a pear-shaped flask to remove the Dowex.
6. Dry the filtrate in a rotary evaporator at $40°C$.
7. Add 1 ml methanol and evaporate to dryness in a rotary evaporator to remove the $NaBH_4$. (Repeat this three times.)
8. Add 1 ml pyridine : acetic anhydride solution (1 : 1) to the dry pear-shaped flask and close the flask with a glass stopper.
9. Place in a boiling water bath for 15 minutes.
10. Add 1 ml water.
11. Dry in the rotary evaporator at $45°C$.
12. Repeat steps 10 and 11.
13. Wash the pear-shaped flask thoroughly three times with 200 μl chloroform (or ethyl acetate) and transfer the solvent containing the sugars into a small tube.
14. Dry in a vacuum desiccator over $CaCl_2$.
15. Seal the tube and store at $-20°C$.
16. Add about 500 μl chloroform (or ethyl acetate) prior to gas chromatographic analysis.

Reference

1. Sawardeker, J. S., J. H. Sloneker and A. R. Jeanes, *Anal. Chem.* **12**, 1602 (1965).

72 PROTEINS (BIURET)

Objective

This is a rapid method for quantitative estimation of proteins but it is about ten times less sensitive than the Lowry method (procedure 73).

Standard

bovine serum albumin 10 $\mu g/1$ μl (1%), linear 1 to 5 μl

Method

1. 500 μl sample.
2. 750 μl biuret reagent solution. *Shake*.
3. Place in a water bath at 37°C for 30 minutes.
4. Read at 540 mμ.

Reagent

Biuret reagent solution is available commercially.

References

1. Herriott, R. M., *Proc. Soc. Exp. Biol. Med.* **46**, 642 (1941).
2. Nowotny, A., *Basic Exercises in Immunochemistry*, p. 99, Springer-Verlag, New York, 1969.

73 PROTEIN (LOWRY)

Objective

This is the classical and one of the most sensitive methods for measuring the concentration of proteins (as little as 0.2 μg). Color is developed with the Folin phenol reagent after alkaline copper treatment. It is ten to twenty times more sensitive than measurement of the ultraviolet absorption at 280 mμ and is ten times more sensitive than the Biuret reaction (procedure 72).

Standard

bovine serum albumin 10 μg/μl, linear 1 to 10 μl

Method

1. 75 μl sample.
2. 750 μl reagent C. *Shake*.
3. Let stand for 10 to 30 minutes.
4. 75 μl Folin phenol reagent D. *Shake*.

5. Let stand for a minimum of 20 minutes (but no longer than 2 hours) at room temperature.
6. Read at 750 mμ.

Reagents

A. 2 g NaOH, 10 g Na_2CO_3, 0.1 g Na-K tartrate per 500 ml water.
B. 0.5 g $CuSO_4 \cdot 5H_2O$ per 100 ml water.
C. Mix 10 ml solution A + 0.2 ml solution B prior to use.
D. Mix 1 part of Folin phenol (2 N) with 1 part water prior to use.

Reference

1. Lowry, O. H., N. J. Rosebrough, A. L. Farr and R. J. Randall, *J. Biol. Chem.* **193**, 265 (1951).

74 RAPID ESTIMATION OF SUGARS BY CELLULOSE THIN LAYER CHROMATOGRAPHY

Objective

This is a method for tentative estimation of sugars. It is advantageous in that it can be run in a short period of time, small amounts of material to be tested can be utilized and acids, e.g., sulfuric acid, can be used as a detection spray as opposed to paper chromatography. (See procedure 50, Estimation of Sugars by Descending Chromatography.)

Standard

0.01 M sugar

Method

1. Sugar preparations (from procedures 17, 18 or 37, etc.).
2. Add 50 μl water.
3. Spot 1 μl at the origin.
4. Fan dry the spots.

5. Place plate in a thin layer tank containing the proper solvents A, B, etc.
6. Dry the plates at room temperature.
7. Stain appropriately. (See procedures 75, 76, 77, etc.).

Reagents (Solvents)

A. n-butanol: pyridine: H_2O (6:4:3 by volume).
B. 1. ethyl acetate: pyridine: acetic acid: water (5:5:1:3 by volume).
   2. pyridine: ethyl acetate: H_2O (11:40:6 by volume). This is used in a beaker in the bottom of the tank to saturate the atmosphere in solvent system B(1).

References

1. Cellulose for Thin Layer Chromatography can be purchased from Macherey, Nagel & Co. as Cellulosepulver MN 300 G (distributed by Brinkman Instruments, Westbury, New York).
2. Fischer, F. G. and H. Dörfel, *Hoppe-Seyler's Z. Physiol. Chem.* **301**, 224 (1955).

75 STAINING OF AMINO SUGARS AND AMINO ACIDS AFTER CHROMATOGRAPHY WITH NINHYDRIN

Objective

This is the classical method of staining amino groups. The amino sugars are blue gray and the amino acids are blue violet after staining with ninhydrin. These colors fade, however, and therefore the spots are fixed with acidic $Cu(NO_3)_2$ which causes all the spots to turn red and remain stable.

Method

1. Use dried thin layer plate or dried paper chromatogram.
2. Spray with ninhydrin aerosol (Nutrition Biochemical Corporation).
3. Dry at room temperature overnight in a hood.

4. Dip chromatogram in reagent A.
5. Let dry at room temperature.

Reagents

A. 1. 5 ml aqueous saturated $Cu(NO_3)_2$.
   2. 0.1 ml HCl (concentrated).
   3. Dilute up to 500 ml with acetone.

Reference

1. Oden, S. and B. Hofsten, *Nature* 173, 449 (1954).

76 STAINING OF SUGARS AFTER CHROMATOGRAPHY WITH ANILINE PHTHALATE

Objective

This is a reagent for detection of reducing carbohydrates but it is not as sensitive as alkaline silver nitrate (procedure 77). It is advantageous, however, in that the reducing sugars have characteristic colors: aldopentose–red, aldohexose–brown, methylpentose–yellow-brown (e.g., fucose, rhamnose, etc.), 4-deoxyhexose, 2-deoxyhexose–orange.

Method

1. Use dried thin layer plate or dried paper chromatogram.
2. Spray with aniline phthalate reagent A.
3. Dry at room temperature.
4. Place in a 100°C oven for 10 minutes.

Reagents

A. 1. 8.3 g phthalic acid.
   2. 4.65 g of distilled aniline.
   3. Dilute to 500 ml with reagent B.

B. 1. In a separatory funnel add butanol to water. *Shake vigorously*.
   2. Let stand until 2 distinct layers have formed.
   3. Use upper water saturated butanol layer.

Reference

1. Partridge, S. M., *Nature* **164**, 443 (1949).

77 STAINING OF THIN LAYER OR PAPER CHROMATOGRAMS FOR SUGARS BY Ag⁺/OH⁻

Objective

All reducing carbohydrates can be detected with this alkaline silver nitrate. After fixation with sodium thiosulfate the reducing sugars appear as stable brown-black, silver containing spots. It is interesting to note that mannose requires 8 to 10 minutes to develop and rhamnose requires 3 to 4 minutes to develop.

Method

1. Chromatograms are stained with silver nitrate reagent A. (This is best achieved by dipping the chromatograms through a trough of reagent A.)
2. Dry at room temperature.
3. Stain with sodium hydroxide reagent B. (This is best achieved by dipping the chromatograms through a trough of reagent B.)
4. After spots have developed, immediately fix with sodium thiosulfate reagent C.
5. Wash with tap water.

Reagents

A. Add 2.5 ml of saturated aqueous $AgNO_3$ to 500 ml of acetone.*
B. Dissolve 16 g NaOH with a small volume of water (to attain a saturated solution) and dilute up to 800 ml with methanol (0.5 N).
C. 1. 50 g $Na_2S_2O_3 \cdot 5H_2O$ dissolved *in* 475 ml water.
   2. Add 475 ml methanol.

*If a precipitate is created add water dropwise until the precipitate disappears.

Reference

1. Trevelyan, W. E., D. P. Procter and J. S. Harrison, *Nature* 166, 444 (1950).

78 STANDARD CURVE FOR MICROANALYTICAL METHODS

Objective

This is a procedure to illustrate how the standard curves are prepared for all the spectrophotometric chemical, quantitative microanalytical methods in this book. This example method uses glucose estimation (procedure 57, D-Glucose) as a model but it is generally applicable. The calibrations are used to determine the concentration of the chemical component or group of components in $\mu g/\mu l$ or as a percentage of the total, e.g., percentage of glucose in the estimated material.

These methods can be easily converted to macromethods by multiplying the amounts given in the procedure by an appropriate factor i.e., generally 4 to 10.

Method

1. To each of 8 tubes add 50 μl of water.
2. From tube 1 remove 1 μl of water, 3 μl from tube 2, 5 μl from tube 3, and 10 μl from tube 4.
3. Add 1 μl of D-glucose standard (2.0 $\mu g/\mu l$) to tube 1, 3 μl to tube 2, 5 μl to tube 3, and 10 μl to tube 4. *Shake*.
4. Tube 5 is a blank which just contains 50 μl of water and then reagents added.
5. From tube 6 remove 1 μl of water, 5 μl from tube 7, and 10 μl from tube 8.
6. Using an appropriately diluted sample to be assayed for D-glucose, add 1 μl of sample to tube 6, 5 μl to tube 7, and 10 μl to tube 8. *Shake*. Usually a 1% solution (5 mg dissolved up to 500 μl with water i.e., 10 $\mu g/\mu l$) of the sample, e.g., lipopolysaccharide, is an appropriate dilution.
7. Proceed with steps 2 through 6 as described in procedure 57, D-Glucose (Glucose Oxidase).

8. Using linear graph paper plot the optical density on the ordinate (y-axis) and the μg on the abscissa (x-axis), i.e., if the standard contains 2 μg/μl and you added 2 μl, then it is 2 μg/μl × 2 μl = 4 μg (Fig. 1).

9. In your unknown samples to be assayed (tubes 6, 7 and 8) you determine the point on the ordinate (y-axis) at which the optical density of your unknowns fall. From the abscissa you may determine the number of μg per 1 μl, 5 μl and 10 μl. By division determine the number of μg/μl in tubes 6, 7 and 8. Average the results of the μg/μl in tubes 6, 7 and 8 and this gives you the final μg/μl in your unknown sample.

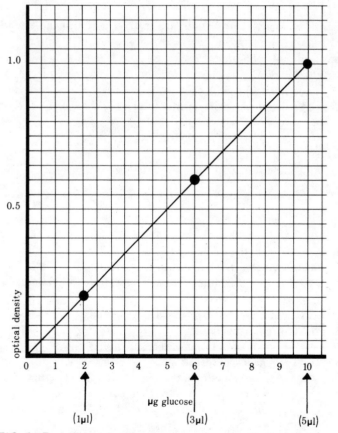

FIG. 1 Experimental calibration curve for quantitation of glucose.

10. However, if you wish to express the concentration of glucose as a percentage of the total estimated material and since the solution you made was 1% (10 $\mu g/\mu l$) and if you determine glucose in this solution to be 2 $\mu g/\mu l$ then you have 20% glucose (2 $\mu g/\mu l$/ 10 $\mu g/\mu l$ × 100) in the unknown material.

79 SUGARS (DISCHE SPECTRUM)

Objective

This is a qualitative and quantitative method useful in the tentative identification of unknown sugars. After heating with sulfuric acid, transformed and split products of sugars are formed with each sugar having characteristic breakdown products demonstrable in the 230 to 700 mμ spectral region. With the addition of cysteine new compounds are formed which yield characteristic spectra which change with time.

Standard

0.001 M sugar sample, 20 μl

Method

1. Add 200 μl sample in an ice bath.
2. Cool to 4°C in the ice bath.
3. Add 900 μl H$_2$SO$_4$ reagent A to the tubes in the ice bath. *Shake*.
4. Let stand for 10 minutes at room temperature (or 3 minutes in a 22 to 25°C water bath).
5. Let stand for 3 minutes in a boiling water bath.
6. Cool to room temperature.
7. Measure the spectrum from 230 to 700 mμ.
8. 20 μl cysteine reagent B. *Shake*.
9. Let stand for 90 minutes in the dark at room temperature.
10. Read the spectrum from 230 to 700 mμ.
11. After 22 hours remeasure the spectrum from 230 to 700 mμ.

Reagents

A. 2 ml water + 12 ml concentrated H_2SO_4.

B. 0.3 g cysteine hydrochloride diluted up to 10 ml with water (3%). Prepare prior to use.

Reference

1. Dische, Z., L. B. Shettles and M. Osnos, *Arch. Biochem. Biophys.* 22, 169 (1949).

80 URONIC ACIDS (QUALITATIVE)

Objective

This is a method to identify the presence or absence of a uronic acid in biological material.

Standards

glucuronic acid 194 mg/10 ml (0.1 M)
galacturonic acid 194 mg/10 ml (0.1 M)

Method

1. Hydrolyze the polysaccharide sample in 1 N H_2SO_4 (approximately 10 mg in 1.0 ml) for 4 hours in a boiling water bath.
2. Neutralize with saturated $Ba(OH)_2$ to pH 7.0.
3. Centrifuge to remove the precipitated $BaSO_4$ (2000 rpm, 10 minutes at 4°C).
4. Dry the supernatant in a vacuum desiccator over $CaCl_2$ (see procedure 12, Drying Samples in a Vacuum Desiccator).
5. Add 40 μl of water.
6. Apply 1 μl standard or sample to the origin on high voltage paper electrophoresis apparatus.
7. Electrophoresis is performed for 90 minutes at 150 ma; 3000 volts using Whatman No. 1; electrophoresis buffer-pyridine: acetic acid: water (10:4:86 by volume) pH 5.3.

8. Air dry the strips of paper.
9. Stain by spraying with naphthoresorcinol reagent B.
10. Air dry the stained strips of paper.
11. Heat the strips of paper for 3 to 4 minutes in an oven at 100°C.

Reagents

A. 0.2 g naphthoresorcinol/100 ml 95% ethanol (0.2%).
B. 1. 18 ml from solution A.
   2. Add 2 ml water.
   3. Add 2 ml H_3PO_4. *Shake.*

Note: Uronic acids will give a blue spot against a pink background.

Reference

1. Partridge, S. M., *Biochem. J.* **42**, 238 (1948).

81 URONIC ACIDS—QUANTITATIVE (DISCHE)

Objective

Uronic acids are primarily constituents of acidic polysaccharides of capsular antigens of bacteria. The method was developed for hexuronic acids. It is necessary to utilize the isologous authentic standard e.g., mannuronic acid standard, to quantitate mannuronic acid in the biological material, to obtain accurate quantitation. Mannuronic acid gives only 17% of the extinction as glucuronic acid.

Standard

standard uronic acid 1 μg/1 μl in water, linear 1 to 10 μl

Method

1. 200 μl sample.
2. Cool to 4°C in an ice bath.
3. Add slowly 1200 μl H_2SO_4 (concentrated) while the tubes are in the ice bath. *Shake.*

4. Heat for 20 minutes in a boiling water bath.
5. Cool to room temperature.
6. Add 40 μl carbazole reagent A. *Shake well.*
7. Allow to stand for 2 hours at room temperature (the color is stable until 3 hours).
8. Read at 535 mμ.

Reagent

A. 10 mg carbazole diluted to 10 ml with 95% ethanol (0.1%). Prepare prior to use.

References

1. Dische, Z. *J. Biol. Chem.* **167**, 189 (1947).
2. Dische, Z. *Methods of Biochemical Analysis*, edited by D. Glick, vol. 2, p. 313, Interscience Publisher Inc., New York, 1955.

3

Biological Characterization

82 COMPLEMENT FIXATION

Objective

The fixation of complement is a serological reaction in which the specific antibody (amboceptor) binds with its homologous antigen and creates a complex which binds (fixes) complement. With a corpuscular antigen, e.g., bacteria or red blood cell, the fixation of complement causes the lysis of the antigen which is a visible reaction. If the antigen is not corpuscular, e.g., cardiolipin, viral, etc., then in order to visualize the fixation of complement, the system must be coupled with a hemolytic (red blood cell) indicator system.

Complement fixation is used if you have a known antigen and wish to determine if its antibody is present. Also, it is used if you have a known antibody and you wish to determine the antigen. If the complement is fixed (i.e., antibody is present to the homologous antigen) then the indicator sheep red blood cells are not lysed.

82A Complement Titration

1. The complement must be titrated each day the test is performed.
2. Add 1 volume of standardized washed sheep red blood cells (2.8%–see procedure 82C) to 1 volume of optimal concentration (see procedure 82D Hemolysin (Amboceptor) Titration) of commercially available antisheep hemolysin.
3. Incubate for 15 minutes at 25°C.
4. Prepare a 1:400 dilution of commercially available complement in buffer reagent B.
5. Into 4 tubes add the following as described in Table 1 below:

Table 1

Reagent in ml	Tube number			
	1	2	3	4
Buffer reagent B	0.6	0.55	0.5	0.4
1:400 complement	0.2	0.25	0.3	0.4
Sensitized red blood cells (step 2)	0.2	0.2	0.2	0.2

6. Incubate tubes for 45 minutes at 37°C in a shaking water bath.
7. Centrifuge the tubes at 2000 rpm for 10 minutes.
8. Read the O.D. of the supernatant at 541 mμ.
9. Compare to the standard curve (procedure 82B, Hemoglobin Standard Curve) and determine graphically on log-log paper (Fig. 1) the 50% hemolysis concentration as shown below:

Table 2

1:400 Dilution of Complement (ml)	% Hemolysis (y) [*]	Ratio $y/100$-y
0.20	30	0.43
0.25	35	0.54
0.30	70	2.33
0.40	85	5.7

[*]y = percent hemolysis obtained from standard curve, Procedure 82B.

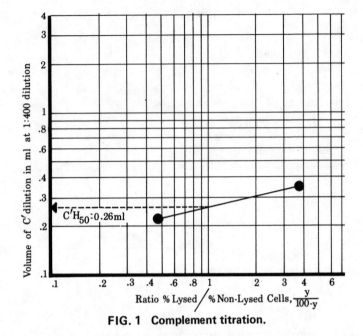

FIG. 1 Complement titration.

From Casey, H. L. Part II. Adaptation of LBCF method to micro technique. In Standardized Diagnostic Complement Fixation Method and Adaptation to Micro Test, Public Health Monograph No. 74, Public Health Service Publication No. 1228. U.S. Government Printing Office, Washington, D.C. 1965.

10. Since 5 $C'H_{50}$ (50% complement units) are required in the diagnostic test (procedure 82F, Complement Fixation Test) and 0.4 ml is utilized in the test, the dilution of complement necessary must be calculated as follows:

$$\begin{array}{l}\text{dilution of complement}\\ \text{necessary to obtain}\\ \text{5 } C'H_{50} \text{ in 0.4 ml}\end{array} = \frac{400 \times 0.4}{5 \times 0.26}$$

Example: $160 \div 1.3 = 123$ and therefore the dilution of complement that should be utilized is $1:123$.

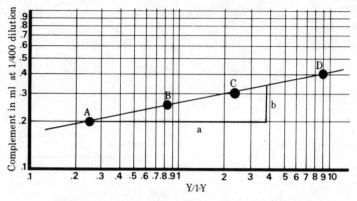

FIG. 2 Determination of complement titration.

From Casey, H. L. Part II. Adaptation of LBCF method to micro technique. In Standardized Diagnostic Complement Fixation Method and Adaptation to Micro Test, Public Health Monograph No. 74, Public Health Service Publication No. 1228. U.S. Government Printing Office, Washington, D.C. 1965.

Note:

In order for the test to work properly the slope of the curve (Fig. 2) must be 0.20 ± 0.02. The slope can be easily obtained by measuring from any point near the left end of the curve 10 cm horizontally to the right. Then measure the vertical distance upward from the right end of the horizontal line to the curve. Divide the vertical measurement by 10 cm to obtain the slope. (Figure 2 has been reduced twofold.)

If the slope is not 0.20 ± 0.02 then the experiment must be repeated until such a result is obtained.

82B Hemoglobin Standard Curve for Complement Fixation

1. Add 1.0 ml of standardized sheep red blood cells (2.8% from procedure 82C, Preparation of Sheep Red Blood Cells) and 7.0 ml of water. *Shake.*
2. After the cells have lysed add 2.0 ml of buffer reagent A.
3. Prepare a 1:10 dilution (0.28%) of sheep red blood cells by adding 1.0 ml of standardized sheep red blood cells (2.8% procedure 82C,

Preparation of Sheep Red Blood Cells), to 9.0 ml of buffer reagent B.

4. Prepare color standards as described below in Table 1:

Table 1

Reagents in ml	Tube number										
	0	1	2	3	4	5	6	7	8	9	10
Hemoglobin solution (step 2)	0	0.1	0.2	0.3	0.4	0.5	0.6	0.7	0.8	0.9	1.0
0.28% cells (step 3)	1.0	0.9	0.8	0.7	0.6	0.5	0.4	0.3	0.2	0.1	0
% hemolysis	0	10	20	30	40	50	60	70	80	90	100

5. Shake the tubes and then centrifuge at 2000 rpm for 10 minutes.
6. Read the O.D. of the supernatant at 541 mμ.
7. Plot a standard linear curve of the % hemolysis versus the O.D. at 541 mμ.

82C Preparation of Sheep Red Blood Cells for Complement Fixation

1. Sheep red blood cells (in Alsevers solution) are washed three times at 2000 rpm with cold buffer reagent B.
2. The packed cells are suspended in buffer reagent B to a concentration of 3% (3 ml packed cells + 97 ml of buffer reagent B).
3. Filter the cells through a funnel containing gauze or absorbant cotton.
4. 1.0 ml of the filtered suspension of cells is lysed with exactly 14 ml of buffer reagent E.
5. Read at 541 mμ in a 1 cm cuvette in a Beckman DU Spectrophotometer (against a buffer reagent E blank).
6. A suspension of 0.452 O.D. is desired (2.8%) and can be obtained using the following formula to determine the total final volume (V_{final}) necessary:

$$V_{final} = V_{initial} \times \frac{O.D._{initial}}{0.452}$$

$V_{initial}$ = initial volume

$O.D._{initial}$ = optical density obtained

V_{final} = the final total volume to which the sheep red
blood cell suspension should be diluted

Example: $100 \times \dfrac{.480}{.452} = 106$ ml final total volume

Therefore 6.0 ml of buffer reagent B is added to the initial 3% solution of sheep red blood cells (100 ml + 6 ml = 106 ml).

Note:

Sheep red blood cells in Alsevers solution are stable at 4°C for one month and therefore once the cells have been standardized the stock cells in Alsevers solution must merely be washed and diluted prior to use.

82D Hemolysin (Amboceptor) Titration

1. 2.0 ml of commercially available glycerinized antisheep hemolysin (amboceptor) is diluted with 98 ml of reagent B (and may be stored in small aliquots at $-20°C$).
2. Add 10 ml of 1:100 hemolysin dilution from step 1 to 90 ml reagent B (dilution 1:1000).
3. Using 6 test tubes dilute the hemolysin as shown in Table 1:

Table 1

Hemolysin		Reagent A	Final Hemolysin Dilution
1. 1 ml of hemolysin from step 1	+	9 ml	1:1000
2. 1 ml of 1:1000 from step 2	+	1 ml	1:2000
3. 1 ml of 1:1000 from step 2	+	1.5 ml	1:2500
4. 1 ml of 1:1000 from step 2	+	2.0 ml	1:3000
5. 1 ml of 1:1000 from step 2	+	3.0 ml	1:4000
6. 1 ml of 1:1000 from step 2	+	7.0 ml	1:8000

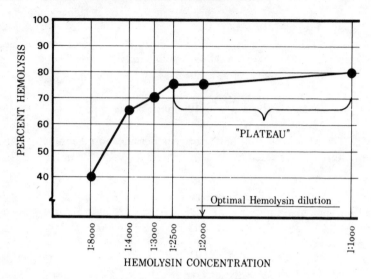

FIG. 3 Hemolysin titration.

From Casey, H. L. Part II. Adaptation of LBCF method to micro technique. In Standardized Diagnostic Complement Fixation Method and
Adaptation to Micro Test, Public Health Monograph No. 74, Public
Health Service Publication No. 1228. U.S. Government Printing Office, Washington, D.C. 1965.

Mix the contents of the tubes well.

4. Add 1.0 ml of the standardized 2.8% sheep red blood cells (procedure 82C, Preparation of Sheep Red Blood Cells) to 1 ml of
   each of the dilutions of hemolysin (from step 3).
5. Incubate for 15 minutes at 25°C in a shaking water bath.
6. Prepare a 1:400 dilution of complement in cold buffer reagent B
   (0.25 ml complement + 99.75 ml cold buffer reagent B).
7. To 6 new tubes add 0.4 ml of cold reagent B, 0.4 ml of the
   1:400 complement dilution (from step 6) and 0.2 ml sheep red
   blood cells sensitized with the hemolysin dilution (from step 5).
8. Incubate in a 37°C water bath with constant shaking for 1 hour.
9. Centrifuge the tubes at 2000 rpm for 10 minutes at 4°C.
10. Read the O.D. of the supernatant at 541 mμ.

11. The amount of hemolysis obtained with the various dilutions of hemolysin is plotted on semi-logarithmic paper (see Fig. 3).[*] Proceeding from left to right the optimal hemolysin dilution is the second dilution on the plateau. In the example shown in Figure 3, the optimal hemolysin dilution is 1:2000.

Note:

This is necessary for determination of the dilution of hemolysin (amboceptor) for optimal sensitization of sheep red blood cells for complement fixation. It is only necessary to titer hemolysin once for each lot.

82E Antigen Titration for Complement Fixation

1. Add to 1.0 ml of your serum to be tested 7.0 ml of buffer reagent B.
2. Incubate in a 56°C water bath for 30 minutes to inactivate any complement in the serum.
3. Add to each of 6 tubes 2.0 ml of buffer reagent B and 2.0 ml of 1:8 dilution of the serum to the first tube and serially dilute the serum.
4. Add to each of 8 tubes 2.0 ml of buffer reagent B and add 2.0 ml of antigen to be tested to the first tube. Serially dilute the antigen.
5. Steps 3 and 4 are stock serum and antigen, respectively, that are used in the dilutions described in Table 1 and Table 2.
6. Add 0.2 ml of the serum dilutions to the appropriate tubes (see Table 1). Add 0.2 ml of antigen to each dilution of serum.
7. Add 0.2 ml of buffer reagent B, rather than antigen, to the serum controls.
8. Shake all the tubes and let stand for 15 minutes at room temperature.
9. Add 0.4 ml of cold complement containing 5 $C'H_{50}$ (see procedure 82A, Complement Titration) to all the tubes in Table 1.
10. Prepare the complement-antigen controls as shown in Table 2.
11. Shake all the tubes and place them in the refrigerator for 15 to 18 hours.

[*]The abscissa of the hemolysin dilution is logarithmic and the ordinate, with the % hemolysis, is linear.

12. Add 0.2 ml of sensitized cells (procedure 82D, Hemolysin [Amboceptor] Titration) to all the tubes; shake and incubate in a 37°C shaking water bath for 30 minutes.

Table 1   Antigen Titration

Antigen Dilution	Antiserum Dilution				
	1:16	1:32	1:64	1:128	1:256
1:2					
1:4					
1:8					
1:16					
1:32					
1:64					
1:128					
1:256					
Serum Control[a]					

[a]The serum control should have 100% hemolysis at each dilution of antiserum because there is no antigen present to fix (i.e., remove) the complement.

Table 2   Complement-Antigen Controls

Antigen dilution[†]	5 $C'H_{50}$[a]			2.5 $C'H_{50}$			1.25 $C'H_{50}$		
	B[‡]	antigen[†] (ml)	C'	B[‡]	antigen[†] (ml)	C'	B[‡]	antigen[†] (ml)	C'
1:2 (tubes 1–3)	0.2	0.2	0.4	0.4	0.2	0.2	0.5	0.2	0.1
1:4 (tubes 4–6)	0.2	0.2	0.4	0.4	0.2	0.2	0.5	0.2	0.1
1:8 (tubes 7–9)	0.2	0.2	0.4	0.4	0.2	0.2	0.5	0.2	0.1
1:16 (tubes 10–12)	0.2	0.2	0.4	0.4	0.2	0.2	0.5	0.2	0.1
1:32 (tubes 13–15)	0.2	0.2	0.4	0.4	0.2	0.2	0.5	0.2	0.1
1:64 (tubes 16–18)	0.2	0.2	0.4	0.4	0.2	0.2	0.5	0.2	0.1
1:128 (tubes 19–21)	0.2	0.2	0.4	0.4	0.2	0.2	0.5	0.2	0.1
1:256 (tubes 22–24)	0.2	0.2	0.4	0.4	0.2	0.2	0.5	0.2	0.1
Controls — Serum diluted 1:16[*] (ml)	0.2	serum* 0.2	0.4	0.4	serum* 0.2	0.2	0.5	serum* 0.2	0.1
Controls — Buffer reagent B[‡] (complement control)	0.4	—	0.4	0.6	—	0.2	0.7	—	0.1

[a]All the 5 $C'H_{50}$ tubes should have 100% hemolysis.

13. Centrifuge at 2000 rpm for 10 minutes and read the O.D. of the supernatants at 541 mμ of the tubes not showing complete lysis.

Note:

Steps 11, 12, and 13 are performed for the reactions in Table 1 and Table 2.

14. Acceptable % hemolysis for complement controls (Table 3) are:

Table 3   Acceptable Percent Hemolysis

Type of Control	5 C'H$_{50}$	2.5 C'H$_{50}$	1.25 C'H$_{50}$
Antigen	100%	85–100%	0–75%
Reagent B	100%	90–100%	40–75%
Serum diluted 1:16	100%	90–100%	0–75%

Evaluation

1. For general purposes the dilution of antigen used is the dilution which gives 30% or less hemolysis and fits in the guidelines of the complement controls (Table 3). The optimal dilution of antigen is the highest dilution which gives the highest complement fixation antibody titer (i.e., greatest amount of complement fixed and therefore 30% or less hemolysis) with the highest dilution of antiserum (see Table 4).

2. Once an antigen lot has been titrated it need not be titrated again.

Table 4   Antigen Titration Evaluation

Antigen Dilution	Antiserum Dilutions						Complement-Antigen Controls		
	1:2	1:4	1:8	1:16	1:32	1:64	5 C'H$_{50}$	2.5 C'H$_{50}$	1.25 C'H$_{50}$
	% of Hemolysis						% of Hemolysis		
1:2	0	0	30	70	100	100	100	70	10
1:4	0	0	10	50	90	100	100	100	40
1:8	0	0	0	(20)	60	90	100	100	50
1:16	0	0	10	40	90	100	100	100	50
1:32	0	0	40	80	100	100	100	100	50
None	100	100	100	—	—	—	—	—	—
							Complement-Reagent B Controls		
							100	100	45

In Table 4 the 1:8 dilution of antigen would be selected as optimal since it is the highest dilution of antigen which gives 30% or less hemolysis with the highest dilution of antiserum.

82F Complement Fixation Test

1. Prepare a 1:4 dilution of the serum to be tested with buffer reagent B (0.1 ml serum + 0.3 ml buffer reagent B).
2. Inactivate the complement in the 1:4 dilution of this serum by incubating in a water bath at 56°C for 30 minutes.
3. Serially dilute the serum by adding 0.2 ml of buffer reagent B to each of 6 tubes. Add 0.2 ml 1:4 diluted serum to the first tube and serially dilute the serum.
4. Add 0.2 ml of the optimal dilution of antigen (see procedure 82E, Antigen Titration) to each of the tubes.
5. Add 0.2 ml of buffer reagent B instead of antigen to the serum controls (see Table 1).
6. Add the indicated volume of buffer reagent B to the complement controls (see Table 1).
7. Prepare a sufficient quantity of the complement dilution to contain 5 $C'H_{50}$ in 0.4 ml (see procedure 82A, Complement Titra-

Table 1

	Serum Dilution ml		Antigen Optimal Dilution ml	Reagent B	Complement ml (5 $C'H_{50}$ in 0.4 ml)	Sensitized Cells
	1:8	1:16→[a]				
Unknown test serum	0.2	0.2	0.2	—	0.4	0.2
Serum-anticomplementary control	0.2	—	—	0.2	0.4	0.2
Positive control serum	0.2	0.2	0.2	—	0.4	0.2
Negative control serum	0.2	0.2	0.2	—	0.4	0.2
Complement Controls						
5 $C'H_{50}$	—	—	0.2	0.2	0.4	0.2
2.5 $C'H_{50}$	—	—	0.2	0.4	0.2	0.2
1.25 $C'H_{50}$	—	—	0.2	0.5	0.1	0.2
5 $C'H_{50}$	—	—	—	0.4	0.4	0.2
2.5 $C'H_{50}$	—	—	—	0.6	0.2	0.2
1.25 $C'H_{50}$	—	—	—	0.7	0.1	0.2
Sheep Cell Control						
Sheep cell control	—	—	—	0.8	—	0.2

[a]For diagnostic tests, dilutions higher than 1:16 are generally unnecessary.

tion, step 10). Add 0.4 ml of 5 $C'H_{50}$ to the assay tubes and the 5 $C'H_{50}$ control. Add 0.2 ml of the 5 $C'H_{50}$ to the 2.5 $C'H_{50}$ control and 0.1 ml to the 1.25 $C'H_{50}$ control tubes.

8. Shake and incubate at 4°C for 16 hours.
9. Prepare a sheep red blood cell control by adding 0.8 ml of buffer reagent B to 0.2 ml of sensitized sheep red blood cells (see procedure 82C, Preparation of Sheep Red Blood Cells).
10. Add 0.2 ml of sensitized sheep red blood cells (procedure 82C, Preparation of Sheep Red Blood Cells) to the remaining tubes.
11. Shake and then incubate in a 37°C water bath for 30 minutes.
12. Centrifuge at 2000 rpm for 10 minutes at 4°C. (In the tubes where there is 100% hemolysis it is not necessary to centrifuge.)
13. Read the O.D. of the supernatant at 541 mμ.

Evaluation

1. The complement controls must fit the guidelines established in Table 3, procedure 82E, Antigen Titration, in order for the test to be valid.
2. If the serum anticomplementary control has less than 75% hemolysis then another serum must be utilized.
3. The end point of serum titrations is the highest dilution showing 30% or less hemolysis. The positive and negative controls are used as a criteria for evaluating the test.

Reagents

A. (Stock Solution)
   1. 85 g NaCl, 3.75 g Na-5,5-diethylbarbiturate. Dissolve in 1400 ml water
   2. Dissolve 5.75 g 5,5-diethylbarbituric acid in 500 ml hot water.
   3. Mix 1 and 2 and cool to room temperature.
   4. Add 5.0 ml reagent C.
   5. Add water to 2000 ml.
   6. Store at 4°C.
B. Prior to use make an accurate five-fold dilution of reagent A with reagent D and the pH must be 7.3 to 7.4.
C. 1. 20.33 g $MgCl_2 \cdot 6H_2O$, 4.40 g $CaCl_2 \cdot 2H_2O$. Dissolve up to 100 ml with water.
   2. Store at 4°C.

D. 1. Add 1.0 g of gelatin to 100 ml water.
   2. Bring to a boil to insure solution of the gelatin.
   3. Make up to 800 ml with water.
   4. Store in the refrigerator not longer than one week.
E. 1. 5.10 g Na-5,5-diethylbarbiturate is dissolved in about 500 ml water.
   2. Add 17.29 ml of 1 N HCl and dilute to 1 liter.
   3. Prior to use dilute five-fold with H_2O (100 ml of (2) + 400 ml water), pH 7.3 to 7.4.

References

1. Casey, H. L., Part II. Adaptation of LBCF method to micro technique. In *Standardized Diagnostic Complement Fixation Method and Adaptation to Micro Test*, Public Health Monograph No. 74, Public Health Service Publication No. 1228. U.S. Government Printing Office, Washington, D.C. 1965.
2. Kabat, E. A. and M. M. Mayer, *Experimental Immunochemistry*, p. 149, Charles C. Thomas, Springfield, Illinois, 1961.

83 Complement Fixation (Bordet-Wassermann)

Objective

This is a diagnostic complement fixation test generally used in Europe. The results are evaluated on the basis of 100% hemolysis of the sensitized indicator cells (as contrasted to the United States where the end point of 50% hemolysis is utilized i.e., $C'H_{50}$).

83A Complement Titration

1. To 1.0 ml of complement add 14.0 ml of buffer reagent B (1:15).
2. To each of 10 tubes add the reagents specified in Table 1.

Table 1

	\multicolumn{10}{c}{Tube number}
	1	2	3	4	5	6	7	8	9	10 (control)
Complement 1:15	0.05	0.10	0.15	0.20	0.25	0.30	0.35	0.40	0.45	—
Buffer reagent B	0.45	0.40	0.35	0.30	0.25	0.20	0.15	0.10	0.05	0.50
Antigen (optimal titer) see procedure 83B	0.50	0.50	0.50	0.50	0.50	0.50	0.50	0.50	0.50	0.50

Table 2

	Tube number									
	1	2	3	4	5	6	7	8	9	10 (control)
Negative control serum	0.10	0.10	0.10	0.10	0.10	0.10	0.10	0.10	0.10	0.10
Complement 1:15	0.05	0.10	0.15	0.20	0.25	0.30	0.35	0.40	0.45	–
Buffer reagent B	0.45	0.40	0.35	0.30	0.25	0.20	0.15	0.10	0.05	0.50
Antigen (optimal titer) see procedure 83B	0.50	0.50	0.50	0.50	0.50	0.50	0.50	0.50	0.50	0.50

3. To an additional 10 tubes a negative control serum is prepared by incubating normal serum at 56°C for 30 minutes (Table 2). Usually Table 2 is utilized for the evaluation of the optimal concentration of complement. Table 1 is only utilized if a negative control serum is unavailable.
4. Incubate in a 37°C shaking water bath for 45 minutes.
5. Add to each tube 0.5 ml of sensitized sheep red blood cells (procedure 83C, Sensitization of Sheep Red Blood Cells). *Shake.*
6. Incubate in a 37°C shaking water bath for 15 minutes.
7. Determine the first tube (Table 2) which shows 100% hemolysis of the sensitized sheep red blood cells. The next tube thereafter is considered the optimal complement concentration, e.g., if the first tube showing 100% hemolysis is tube number 5, then tube number 6 is considered the optimal complement concentration.
8. In order for the test to be valid the complement control (tube 10) must be negative.
9. Complement is titered on each day the tests are run.

83B Antigen Titration

Objective

The purpose of the titration is to determine the optimal concentration of antigen. This titration must only be performed once with each antigen examined.
1. Incubate 2.5 ml of a positive serum in a 56°C water bath for 30 minutes.
2. From the positive serum prepare the following dilutions:
   a. 0.5 ml serum + 0.5 ml buffer reagent B (1:1)

    b. 0.3 ml serum + 1.2 ml buffer reagent B (1 : 5)
    c. 0.5 ml of (b) + 0.5 ml buffer reagent B (1 : 10)
    d. 0.5 ml of (c) + 0.5 ml buffer reagent B (1 : 20)
    e. 0.5 ml of (d) + 0.5 ml buffer reagent B (1 : 40)
    f. 0.5 ml of (e) + 0.5 ml buffer reagent B (1 : 80)

3. Prepare the following dilutions of the antigen to be tested:
    a. 0.2 ml of antigen + 19.8 ml buffer reagent B (1 : 100)
    b. 5.0 ml of (a) + 5.0 ml buffer reagent B (1 : 200)
    c. 2.5 ml of (a) + 5.0 ml buffer reagent B (1 : 300)
    d. 5.0 ml of (b) + 5.0 ml buffer reagent B (1 : 400)
    e. 1.0 ml of (a) + 4.0 ml buffer reagent B (1 : 500)

4. To each of 30 clean tubes add 0.1 ml of the appropriately diluted positive serum (see Table 1).

5. To all of the tubes add the optimal concentration of complement (see procedure 83A, Determination of the Optimal Concentration of Complement) so that the total volume of complement is 0.50 ml, e.g., 0.20 ml complement + 0.30 ml of buffer reagent B.

6. To each of the 6 serum dilutions add 0.5 ml of the appropriate dilution of antigen (see Table 1). *Shake.*

7. Incubate in a 37°C shaking water bath for 45 minutes.

8. Add 0.5 ml of sensitized sheep red blood cells (see procedure 83C, Sensitization of Sheep Red Blood Cells).

9. Incubate in a 37°C shaking water bath for 15 minutes.

Table 1

Positive serum	Antigen				
	1 : 100	1 : 200	1 : 300	1 : 400	1 : 500
1 : 1	+++	+++	+++	+++	+++
1 : 5	+++	+++	+++	+++	+++
1 : 10	++	+++	+++	+++	+++
1 : 20	--	+	+++	+++	+++
1 : 40	--	--	(++)	+	--
1 : 80	--	--	--	--	--

Evaluation

1. Use the highest dilution of antigen which with the highest dilution of positive serum still gives a positive result (see circled dilution in Table 1, 1 : 300). A positive result is considered a tube which still has visible unhemolyzed sheep red blood cells.

83C Sensitization of Sheep Red Blood Cells for Complement Fixation (Bordet-Wassermann)

1. Wash the sheep red blood cells four times with 3 volumes of buffer reagent B by centrifugation at 2000 rpm for 15 minutes at 4°C.
2. Suspend 12.5 ml of packed sheep red blood cells with 487.5 ml of buffer reagent B (2.5%). Store in the refrigerator until ready for use.*
3. Prior to the analysis mix 50 ml of the sheep red blood cell suspension with 50 ml of the prescribed dilution of hemolysin (see step 8, procedure 83D, Hemolysin (Amboceptor) Titration).
4. Incubate in a 37°C shaking water bath for 30 minutes.

83D Hemolysin (Amboceptor) Titration

Objective

This is a method for attaching the hemolysin to the sheep red blood cells, i.e., sensitization.

1. To each of 12 tubes add 0.5 ml of 2.5% sheep red blood cells (step 2, procedure 83C, Sensitization of Sheep Red Blood Cells for Complement Fixation).
2. Add 0.5 ml of a 1:15 dilution of complement to each tube.
3. Dilute 20 µl of hemolysin up to 20 ml with buffer reagent B (1:1000).
4. To 12 clean tubes add the hemolysin concentrations shown in Table 1.
5. Remove 0.5 ml from each hemolysin dilution (step 4) and add it to the tubes in step 2. *Shake.*
6. Incubate in a 37°C shaking water bath for 30 minutes.
7. The *unit* dilution of hemolysin is the highest dilution which gives a complete hemolysis of the sheep red blood cells. The control from Table 1, (l.), should have no hemolysis.
8. For the test (Complement Fixation Test 83E) 2 units of hemolysin are used, e.g., if 1:4000 = 1 unit therefore in the test 1:2000 = 2 units. Two units are used in the complement (83A) and antigen (83B) titration in addition to the complement fixation test.

---

*The sheep red blood cells should be shaken prior to use since the red cells settle on the bottom of the container.

Table 1

Hemolysin (1:1000, from step 3)	Buffer reagent B	Add 0.5 ml from tube shown	Total volume, ml	Dilution
a. 1.0	0.00		1.00	1:1000
b. 0.5	0.25		0.75	1:1500
c. 0.5	0.50		1.00	1:2000
d. 1.0	2.00 >		2.00	1:3000
e. 0.5	1.50 >		1.50	1:4000
f. 0.5	2.00 >	(g)→(d)	2.00	1:5000
g. 0.0	0.50 ←	(f)→(d)	1.00	1:6000
h. 0.25	1.50	(e)→(i)	1.75	1:7000
i. 0.0	0.50 ←	(f)→(k)	1.00	1:8000
j. 0.0	1.00 ←		1.50	1:9000
k. 0.0	0.50 ←		1.00	1:10,000
l. 0.0	1.00		1.00	Control

9. The hemolysin must be titered only once for each lot of hemolysin utilized. The hemolysin can be purchased commercially.

83E Diagnostic Complement Fixation (Bordet-Wassermann)

1. Incubate all sera to be tested in a 56°C water bath for 30 minutes (inactivation of complement).
2. Add 0.1 ml of each serum to test tubes.
3. Add 0.1 ml of an inactivated (step 1) positive control serum and 0.1 ml of an inactivated (step 1) negative control serum to 2 additional tubes.
4. Add the appropriate volume of buffer reagent B so that with the optimal concentration of complement the final volume of comple-

ment and buffer reagent B is 0.5 ml (see procedure 83A, Complement Titration).

5. Add 0.5 ml of the optimal antigen dilution (see procedure 83B, Antigen Titration). *Shake*.
6. Incubate in a 37°C shaking water bath for 45 minutes.
7. Add 0.5 ml of sensitized sheep red blood cells (procedure 83D, Hemolysin [Amboceptor] Titration). *Shake*.
8. Incubate in a 37°C shaking water bath for 15 minutes.
9. Prepare a serum, complement and sheep red blood cell control as shown in Table 1.

Evaluation

1. Control (Table 1)–The antigen control and the complement control must show 100% hemolysis. In the sheep red blood cell control there should be no hemolysis.
2. The positive control serum should show no hemolysis and the negative control serum should show 100% hemolysis.
3. In the tested sera if there are complement fixation antibodies present there will be no hemolysis. However, if complement fixation antibodies are not present in the sera to be tested there will be 100% hemolysis of the sheep red blood cells.

Table 1

	Test	Antigen Control	Negative Serum Control (Inactivated)	Complement Control	Sheep Red Blood Cell Control
Serum (inactivated)	0.1	—	0.1	—	—
Buffer reagent	0.2	0.3	0.7	0.8	1.1
Complement[a]	0.3	0.3	0.3	0.3	—
Antigen	0.5	0.5	—	—	—
	Incubate in a 37°C shaking water bath for 45 minutes.				
Sensitized Sheep Red Blood Cells	0.5	0.5	0.5	0.5	0.5
	Incubate in a 37°C shaking water bath for 15 minutes.				

[a]Assume for this example that the optimal concentration of complement is 0.30 ml (of a 1:15 solution, see procedure 83A, Complement).

Reagents

The buffers (A to D) are prepared as described in Procedure 82.

Reference

1. Kolmer, J. A. and F. Boerner, *Approved Laboratory Technic*, p. 661, Appleton-Century, Inc., New York, 1945.

84 DETECTION OF ANTIBODY-FORMING CELLS TO LIPOPOLYSACCHARIDE (O ANTIGENS)

Objective

This is a method to enumerate spleen or lymph node cells producing antilipopolysaccharide antibodies. Generally the maximum primary immune response occurs from day 3 to 5.

Method

1. Inject mice with 0.5 ml of lipopolysaccharide in saline (100 μg/ml). (Lipopolysaccharide is prepared as described in procedure 27.)
2. 500 μg of lipopolysaccharide in 1 ml saline is incubated with 4.5 ml of 0.25 N NaOH for 2 hours at 37°C.
3. Neutralize with 4.5 ml of 0.25 N HCl.
4. Incubate a 1% suspension of thrice washed sheep red blood cells (see procedure 92, step 2 for washing instructions) with an equal volume of hydrolyzed and neutralized lipopolysaccharide (step 3) at a concentration of 10 μg/ml at 37°C for 30 minutes.
5. The cells are washed three times with 10 volumes of saline and antigen treated erythrocytes are made up to a 4.0×10^9 cells/ml (~20%) suspension in saline.
6. The method is continued as described from step 3, procedure 92, Localized Hemolysis in Gel for Detecting Antibody-forming Cells.

References

1. Landy, M., R. J. Trapani and W. R. Clark. *Amer. J. Hyg.* 62, 54 (1955).

2. Landy, M., R. P. Sanderson and A. L. Jackson, *J. Exp. Med.* **122**, 483 (1965).

3. Koros, A. M., T. R. Roberts and J. Verkleeren, unpublished, 1969.

85 DETERMINATION OF COMMON ANTIGEN BY HEMAGGLUTINATION

Objective

In preparing the common antigen (procedure 6) it is useful to check the purity of the preparation. The purity of the common antigen preparation is determined by passive hemagglutination with antisera prepared in rabbits against heat-killed *E. coli* 014.

Method

1. Prepare a 2.5% sheep erythrocyte suspension in hemagglutination buffer, pH 7.3 (Difco).
2. Wash the cells three times in the buffer (2000 rpm, 10 minutes at 4°C).
3. Suspend the final pellet of erythrocytes to 2.5% in common antigen from step 9, procedure 6 (Common Antigen).
4. Incubate for 30 minutes in a 37°C water bath.
5. Wash the cells three times with hemagglutination buffer (2000 rpm, 10 minutes at 4°C).
6. Antiserum (prepared from heat-killed cell suspensions of *E. coli* 014 injected into rabbits as described in procedure 100, Preparation of O Antigen for Antisera Production, and procedure 98, Preparation of Antisera) was two-fold serially diluted as follows.
7. To each of 12 tubes add 0.1 ml of hemagglutination buffer, pH 7.3 (Difco) and to the first tube add 0.1 ml of antiserum. *Shake*.
8. Serially dilute by removing 0.1 ml from the first tube and successively adding 0.1 ml to each of the next tubes. *Shake*.
9. Add 0.1 ml of coated and washed 2.5% sheep erythrocytes (step 5) to each of the tubes. *Shake*.
10. Incubate in a 37°C water bath for 30 minutes.
11. Centrifuge at 2000 rpm for 10 minutes at 4°C.
12. Read the hemagglutination titer. The titer is the reciprocal of the highest dilution which gives a visible agglutination.

Evaluation

A hemagglutination titer of a relatively pure common antigen preparation should be 512 or more.

Reference

1. Suzuki, T., E. A. Gorzynski and E. Neter, *J. Bacteriol.* **88**, 1240 (1964).

86 DOUBLE DIFFUSION IN ONE DIMENSION (OAKLEY)

Objective

This is a one dimensional double diffusion method in agar. Diffusion of antigens and antibodies from opposite directions results in the formation of multi-layered precipitations in a central column if heterogeneous antigen solutions and suitable antiserum are used. It is likely that the number of lines produced in the diffusion column is the minimum number of antigens present in the diffusing antigenic mixture. This method can be used to demonstrate the complexity of antigenic mixtures, quantitate antigen and follow purification of antigenic components.

Method

1. The test is carried out in capillary tubes 100 by 6 mm which have been internally coated with a film of agar. (See procedure 103, step 2 Simple Diffusion [Oudin].)
2. Add 400 μl of antiserum or antiserum dilutions to a separate test tube.
3. Immediately prior to addition to the capillary tubes, add 400 μl agar reagent A to each of the antisera containing tubes. (Both antiserum and agar should be about 45°C prior to mixing.)
4. Dispense 400 μl of antisera-agar mixture to the capillary tubes (which comes to a height of about 30 mm) with care to avoid air bubbles.
5. Allow to solidify at room temperature for about 4 hours.

6. Add to each tube (with care to avoid air bubbles) 300 μl agar reagent C. (This is about a 22mm high agar layer.)
7. Allow to solidify and leave at room temperature for 2 hours.
8. Add to the capillary tubes 400 μl of two-fold serially diluted antigen. (The antigen should be serially diluted in reagent B.)
9. Seal the tubes with rubber caps or rubber stoppers and melted paraffin around the stoppers to prevent evaporation.
10. Incubate the tubes for 24 to 48 hours at 37°C in a vertical position.
11. Visualization of the precipitate is facilitated by allowing a strong light to pass through the tubes obliquely.

Reagents

A. 2 g agar (ionagar)–dilute up to 100 ml with reagent B. Autoclave for 15 minutes at 15 lbs of pressure (121°C). Add 20 mg merthiolate.
B. 1. 0.12 g diethylbarbituric acid
   2. 0.08 g sodium 5,5-diethylbarbiturate
   3. 1.7 g NaCl
   4. Dilute up to 200 ml with water, pH 7.4.
C. 1 g agar (ionagar) and dilute up to 100 ml with reagent B. Autoclave for 15 minutes at 15 lbs of pressure (121°C). Add 20 mg merthiolate.

Reference

1. Oakley, C. L. and A. J. Fulthorpe, *J. Pathol. Bacteriol.* **65**, 49 (1953).

87 FIXATION AND STAINING OF IMMUNOELECTROPHORETIC OR OUCHTERLONY GEL DIFFUSION PATTERNS

Objective

Staining of the precipitin patterns of immunoelectrophoretic (procedure 88) or Ouchterlony gel diffusion (procedure 94) is of great value for the chemical characterization of the precipitin lines, or for the recording of especially fine precipitin lines.

Method

1. The agar slides or petri dishes are washed with saline for 72 hours with 3 changes of saline daily.
2. The slides or dishes are covered with filter paper and fan dried at room temperature (or for 16 hours in an oven at 37°C).
3. Remove the filter paper and place the slides or dishes in 2% (volume: volume) acetic acid solution for approximately 5 minutes.
4. The slides or dishes are then ready for staining.* (Only 2 of the protein staining methods shall be enumerated.)

Amido black

1. 0.5 g of Amido black (Amido Schwarz 10B) is dissolved in 100 ml reagent A.
2. Stain with this solution for 10 minutes.
3. Wash with reagent A (with several changes of the reagent) for 15 minutes.

Azocarmine

1. 0.5 g of Azocarmine B is dissolved in 100 ml reagent A.
2. Stain with this solution for 15 minutes.
3. Wash with reagent A (with several changes of the reagent) for 15 minutes.

Reagents

A. 450 ml methanol
50 ml glacial acetic acid

Reference

1. Grabar, P. and P. Burtin, *Immunoelectrophoretic Analysis*, p. 30, Elsevier Publishing Co., New York, 1964.

*The slides can be photographed prior to staining by direct contact photography.

88 IMMUNOELECTROPHORESIS

Objective

This procedure utilizes a combination of physiochemical and immunochemical techniques. The material to be examined is separated electrophoretically on agar gel and then exposed to the precipitating antiserum. Both the antibodies and antigens diffuse toward one another, forming, upon confluence, well-defined precipitin lines.

Method

1. 2 g purified agar (ionagar) is dissolved in 50 ml buffer reagent A and 50 ml water by heating in a boiling water bath.
2. Autoclave for 15 minutes at 15 lbs pressure (121°C).
3. Centrifuge the hot agar for 3 minutes at 3000 rpm and discard any undissolved pellet.
4. 10 mg of merthiolate (sodium salt) are added as a preservative.
5. Clean microscope slides with 95% ethanol and dry.
6. 3 ml of agar is placed on each slide.
7. After the agar solidifies the agar-coated slides are placed in a moist chamber at room temperature. (The moist chamber can be a petri dish with water, in a sealed container.)
8. Holes and troughs are punched through the agar layer with a commercially available die (as shown in Fig. 1).
9. To each hole add 1 μl of antigen. Since each slide has 2 holes the

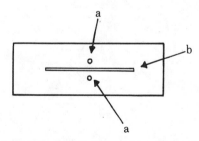

FIG. 1 Immunoelectrophoresis slide, a—holes, b—trough.

same antigen may be added to each hole or a different antigen may be added to the holes. However, enzymatically cleaved antigens which diffuse rapidly are placed only in 1 hole because migration to the opposite side of the trough is possible.

10. The buffer compartments are filled with buffer reagent A.
11. Place the agar slides on an immunoelectrophoresis apparatus.
12. Filter paper strips are used to establish contact between the agar slides and the buffer in the buffer compartments of the immunoelectrophoresis apparatus.
13. Cover the apparatus with a lid to prevent desiccation.
14. The voltage between the ends of the slides is adjusted to 45 V (6 volts/cm) and the electrophoresis is complete in about 45 minutes. (A lid on the apparatus is necessary to prevent loss of moisture due to heating.)
15. After the electrophoresis the agar in the pre-cut center trough is removed with a 19 gauge needle.
16. Add 50 μl of antisera into the trough.
17. The slides are placed into a moist chamber for 20 to 28 hours to obtain optimal immunoprecipitates.

Reagent

A. 6.69 g sodium 5,5-diethylbarbiturate
   4.42 g sodium acetate · 3H$_2$O
   Dissolve up to 750 ml with water. The pH is adjusted to 8.2 by the addition of about 90 ml of 0.1 N HCl.

Reference

1. Grabar, P. and P. Burtin, *Immunoelectrophoretic Analysis*, p. 25, Elsevier Publishing Co., New York, 1964.

89 INHIBITION OF PASSIVE HEMAGGLUTINATION

Objective

The hemagglutination of a system of antigen-coated erythrocytes and antisera can be inhibited by the addition of a soluble antigen or other serologically active compounds to this system. If the soluble antigen reacts specifically with the antisera then there occurs a competition between the reactive groups of the antigen-coated erythrocyte* and the soluble antigen and thus, a competitive inhibition of the hemagglutinating reaction, occurs. This reaction can be utilized for proof of cross reactivity of antigens or antigenic contamination of biological material. It is also useful in assaying for oligosaccharides in blood group antigens, bacterial antigens, natural products and proteins, e.g., chorionic gonadotropic hormone for pregnancy tests.

Method

1. Prepare a 10% suspension of isolated material (e.g., Isolation of Monosaccharides, Oligosaccharides, or Amino Sugars from Paper-Chromatograms, procedure 19) in saline.
2. Add 0.2 ml of saline into 16 tubes.
3. Add 0.2 ml of isolated material (step 1) to the first tube and serially dilute the isolated material.
4. Using the serum from 3 dilutions higher than the titer obtained, i.e., three dilutions above the minimum concentration required for hemagglutination (from procedure 95, Passive Hemagglutination of Lipopolysaccharide or Other Material), 0.2 ml of the serum is added to each tube. *Shake.*
5. The tubes are incubated in a 37°C water bath for 30 minutes.
6. Add to each tube 0.2 ml of the coated sheep red blood cells (Section B, step 15, procedure 95). *Shake.*
7. Incubate in a 37°C water bath for 30 minutes.
8. Read the agglutination. The inhibition titer is considered the reciprocal of the highest dilution of the soluble antigen which inhibits the hemagglutination.

Note: Two control tubes are utilized. In 1 control tube is 0.4 ml of saline and after 30 minutes incubation at 37°C (step 5), 0.2 ml of coated sheep red blood cells are added (step 6) and the procedure continued as described. This tube should have no hemagglutination, i.e., negative control.

A second tube contains 0.2 ml of saline, 0.2 ml of antisera (step 4) and continue the procedure as described. This is the positive control and should have hemagglutination.

---

*There are many methods for coating antigens to red blood cells, e.g., tannic acid method, carbodiimide, bis diazobenzidine, etc.

Reference

1. Kabat, E. A. and M. M. Mayer, *Experimental Immunochemistry*, p. 127, 2d ed., Charles C. Thomas Publisher, Springfield, 1961.

90 LD$_{50}$

Objective

This is a method for measurement and evaluation of an end point in biological procedures at which 50% of test animals react or die.

Method

1. Prepare two-fold serial dilutions of the solution (a convenient stock solution is about 80 mg/ml) in phosphate-buffered saline, pH 7.2.
2. For every dilution inject each of 8 mice with 0.1 ml intravenously.
3. Count the surviving mice at three days. (It is usually convenient to remove the dead mice after 8, 24, 48 and 72 hours.)
4. $\log \text{LD}_{50} = \log D_n + \dfrac{50 - \% \text{ of death at } D_n}{\% \text{ of death at } D_v - \% \text{ of death at } D_n} \times (\log \text{ dilution factor})$

D_n = dilution where the % death is immediately less than 50%
D_v = dilution where the % death is immediately greater than 50%

Sample Calculation

Concentration mg/0.1 ml	Alive	Dead	Total Alive	Total Dead	Total	% Dead
0.5	8	0	22	0	22	0%
1	6	2	14	2	16	12.5%
2	5	3	8	5	13	38.5%
4	3	5	3	10	13	77%
8	0	8	0	18	18	100%

$$\log \text{LD}_{50} = \log 2 + \frac{50 - 38.5}{77 - 38.5} \quad (.3010)$$

$$= .3010 + \frac{11.5}{38.5} \quad (0.3010)$$

$\log \text{LD}_{50} = .3906$

$\text{LD}_{50} = 2.46 \text{ mg}/0.1 \text{ ml}$

Reference

1. Reed, L. J. and H. Muench, *Amer. J. Hyg.* 27, 493 (1938).

91 LD$_{50}$ WITH CYCLOHEXIMIDE PRETREATMENT

Objective

Cycloheximide is an antibiotic which causes a transient interruption of protein synthesis. The lethal effect of the substance tested is potentiated by cycloheximide pretreatment. This method is generally used to compare with the standard LD$_{50}$ (procedure 90). Other investigators have used 37°C pretreatment and maintenance of the animals rather than cycloheximide pretreatment to enhance the lethal effects of the preparation to be tested.

Method

1. Inject half the mice intraperitoneally with 80 mg/kg* in 1.0 ml with cycloheximide.
2. Inject the other half of the mice with 1 ml saline intraperitoneally.
3. After 1 hour, inject 0.1 ml intravenously of two-fold serial dilutions of the solution (a convenient stock solution is about 80 mg/ml) to be tested, into the mice.
4. For every dilution inject 8 mice.
5. Count the surviving mice at three days. (It is usually convenient to remove the dead mice after 8, 24, 48 and 72 hours.)
6. Calculate the LD$_{50}$ of the cycloheximide pretreated and saline-pretreated mice (see procedure 90, LD$_{50}$).

Reagent

*If the mice weigh 20 g then the cycloheximide solution should be 1.6 g/liter saline.

Reference

1. Reed, L. J. and H. Muench, *Amer. J. Hyg.* 27, 493 (1938).

92 LOCALIZED HEMOLYSIS IN GEL FOR DETECTING ANTIBODY-FORMING CELLS (JERNE)

Objective

This is a method for enumerating the cells producing antibodies to sheep red blood cells (or other erythrocytes used as an antigen) in organs such as spleen and lymph nodes. As described, it is a method for measuring primarily IgM (19S) antibody. On successive days after giving a primary injection of sheep red blood cells, lymphoid cells from the animal are mixed with sheep red blood cells localized in agar gel in a petri dish and incubated at 37°C for 1 hour. Single antibody releasing cells form plaques that become visible after the addition of complement.

Method

1. The sheep red blood cells are purchased in Alsever's solution. Sheep red blood cells used as antigen are utilized within one or two days after their arrival. They are washed two to three times by centrifugation at 2000 rpm for 15 minutes with buffer reagent A prior to injection.
2. Sheep red blood cells used for the localized hemolysis in gel are aged at 4°C for one to two weeks. Prior to use, they are washed two to three times with basal medium, Eagle and adjusted to 4.0×10^9 cells/ml.
3. The suspension of spleen cells is prepared by teasing a spleen against a 40 mesh stainless steel wire screen into basal medium, Eagle. The suspension is dispersed with a Pasteur pipette, then filtered through one layer of Curity (grade 90) cheesecloth.*
4. If a carbon dioxide incubator is unavailable the basal medium,

Eagle, is adjusted to pH 7.2 by titration with 0.2 M tris (hydroxymethyl) aminomethane (and thus the need for a carbon dioxide incubator is eliminated).

5. A bottom base layer of 1.4% agar is prepared by mixing equal volumes of melted 2.8% agar with double concentration basal medium, Eagle, both equilibrated to 45°C, and adding 20 ml to 100 X 15 mm Optilux petri dishes. The agar is allowed to cool and harden at room temperature; the petri dishes are inverted and placed without lids into a 37°C incubator for 90 minutes to evaporate excess moisture. The plates can be stored inverted with lids at 4°C for up to seven days. Prior to use, they are equilibrated at room temperature for several hours.

6. A top layer, prepared on the day of assay, of 0.7% agar containing DEAE-dextran is made by mixing equal volumes of autoclaved 1.4% agar with double concentration basal medium, Eagle. Both are equilibrated to 45°C before mixing and kept at this temperature. Immediately prior to assay 0.1 ml DEAE-dextran reagent C is added to 2.4 ml of the agar solutions in 12 X 75 mm plastic disposable test tubes. The solutions in the test tubes are maintained at 45°C in a water bath.

7. To each tube is added in rapid succession, 0.1 ml (4.0 X 10⁹ cells) of aged sheep red blood cells (step 2), and 0.5 ml suspension of spleen cells.

8. The contents of the tubes are mixed and then poured onto the prepared bottom-layered petri dishes. The petri dishes are shaken to obtain an even top layer and the agar solution is allowed to solidify at room temperature.

9. The plates are incubated upright with lids for 1 hour at 37°C.

10. 2 ml of guinea pig complement, diluted ten-fold in basal medium, Eagle, is added and the incubation continued for an additional 30 minutes.

11. The plates are left at room temperature for 2 hours after which complement is poured off and the plates washed twice with saline. They are stored inverted with lids at 4°C overnight.

12. The plates are stained with 8 ml freshly prepared benzidine reagent B for 10 minutes.

13. After pouring off the stain, the plates are washed twice with saline.

14. The antibody-forming cells are counted under low magnification and appear like stars on a dark blue background (Fig. 1).

FIG. 1 A benzidine-stained plate with circular areas of lysis which signify the position of antibody-forming cells.

Reagents

A. Phosphate-buffered saline, pH 7.2
   1. 21.3 g Na_2HPO_4 diluted to 1 liter with water (0.15 M).
   2. 10.2 g KH_2PO_4 diluted to 500 ml with water (0.15 M).
   3. 8.77 g NaCl diluted to 1 liter with water (0.15 M). Add 286 ml
      (1) + 90 ml (2) + 376 ml (3). Adjust the pH to 7.2 with (1) or
      (2) if necessary.
B. 0.2 g benzidine is dissolved in 10 ml glacial acetic acid to which is
   added 90 ml water containing 0.5 ml 30% H_2O_2.
C. 100 mg DEAE dextran diluted up to 10 ml with basal medium,
   Eagle (1X) adjusted to pH 7.2 with 0.2 M tris (hydroxymethyl)
   aminomethane.

*D. 1 g of crystal violet is dissolved in 1 l of 0.1 M citric acid. Filter. An aliquot of spleen cells in the suspension can be stained with this solution and counted in a hemocytometer if the results are to be expressed per 10^6 spleen cells.

Note:

The peak antibody-forming cell response occurs at day 4 to 5 and the optimal dose of sheep red blood cells (injected either intravenously or intraperitoneally) is 4.0×10^8 (generally in 0.5 ml or less). This schedule has been demonstrated in rats and mice.

References

1. Jerne, N. K., A. A. Nordin and C. Henry, in *Cell Bound Antibodies*, edited by B. Amos and H. Koprowski, p. 109, Wistar Institute Press, Philadelphia, 1963.
2. Jerne, N. K. and A. A. Nordin, *Science* **140**, 405 (1963).
3. Lederer, W. H. Unpublished results, (1970).

93 MICROPRECIPITATION INHIBITION

Objective

If nonprecipitating split products of a polysaccharide are added to a system of precipitating antibodies and polysaccharide antigen, the nonprecipitating split products may interfere with the precipitation reaction. With increased concentration of an actively competing nonprecipitating product (oligosaccharide or monosaccharide), the specific precipitation is more and more inhibited. Thus, by this procedure, one can determine the immunodeterminant structures of a polysaccharide.

Method

1. To the first of 10 tubes add 50 μl of a 1% solution of the mono-or oligosaccharide (procedure 19, Isolation of Monosaccharides, Oligosaccharides, or Amino Sugars from Paper Chromatograms).
2. Prepare two-fold serial dilutions of the inhibitor antigen from step 1, i.e., add 50 μl of saline to each tube and transfer 50 μl from the first tube to each successive tube.

3. Add 25 μl of the centrifuged antiserum used in step 1, procedure 102, Quantitative Microprecipitation, (against the polyvalent polysaccharide antigen, procedure 98, Preparation of Antisera) to each tube.
4. Shake in a 37°C water bath for 30 minutes.
5. Add to each tube 50 μl of the concentration of antigen found in procedure 102, Quantitative Microprecipitation, to be at the maximum of the equivalence zone.
6. Two sets of control tubes are included, one containing antiserum and antigen in appropriate total volume and serum control containing neither antigen nor inhibitor.
7. Shake in a 37°C water bath for 60 minutes.
8. Allow to stand for seven days at 2°C. *Shake gently twice daily.*
9. Continue as described in step 8–14, procedure 102, Quantitative Microprecipitation.

Evaluation

1. From the determined amount of protein in the precipitate, the percentage of the inhibition for the antigen antibody reaction (as compared to procedure 102, Quantititative Microprecipitation) is calculated for each dilution of the mono- or oligosaccharide inhibitor. The percentage inhibition is plotted against the concentration of inhibitor. Inhibitors may be compared at the 50% inhibition.

References

1. Heidelberger, M. and F. E. Kendall, *J. Exp. Med.* **62**, 697 (1935).
2. Kabat, E. A. and M. M. Mayer, *Experimental Immunochemistry*, p. 78, 2d ed., Charles C. Thomas, Springfield, 1961.

94 OUCHTERLONY GEL DIFFUSION

Objective

This is a method of double diffusion in agar. Both antigen and antibody migrate simultaneously into the agar layer between them creating precipitin bands. These bands remain stationary and their width is dependent on the concentration of the components. Therefore, it is semi-quantitative and also permits detection of trace amounts of proteins. The patterns of the precipitin lines created are indicative of the specificity of the antibody (or antigen).

Method

1. 4 g of purified agar (ionagar) are mixed with 100 ml phosphate buffer reagent A and 300 ml of water.
2. The mixture is heated in a boiling water bath until the solution is clear. Autoclave for 15 minutes at 15 lbs of pressure (121°C).
3. Add 40 mg merthiolate as preservative.
4. Centrifuge the hot agar solution for 3 minutes at 3000 rpm to remove any undissolved particles.
5. 22 ml of the hot agar solution are pipetted into an 8 cm diameter petri dish.
6. Allow the agar to solidify.
7. The holes are cut into the agar by a commercially available puncher. (The hole puncher should be dipped into melted, hot paraffin and then left to cool. This cooling of the hole puncher with paraffin prevents the agar from sticking to the hole puncher. The holes should be 0.6 to 0.7 cm in diameter approximately 2 cm apart as measured from the center of the holes, Fig. 1).
8. To 3 ml of the hot melted agar add 6 ml of hot water. Allow this solution to cool to 45°C and add 3 drops with a Pasteur pipette into each hole.

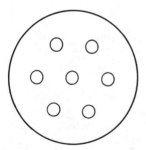

FIG. 1 Pattern of punched holes in Ouchterlony agar gel diffusion plates.

9. Add to each hole in the periphery, 100 μl of a solution of antigen (0.1 to 1%).
10. To the center hole add 100 μl of antisera.

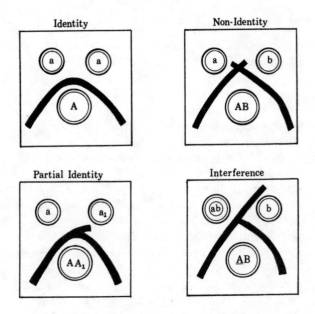

FIG. 2 Basic patterns in diffusion systems.

Code to Fig. 2

1. Antigen "a" and corresponding antibody "A" at equivalent concentrations.
2. "A"–antisera to antigen "a."
3. "B"–antisera to antigen "b."
4. "a_1"–cross reacting antigen to "a," i.e., antigenic factor related to "a."
5. "A" and "A_1"–antibodies of different grades of specificity.
6. " (ab) "–encircled "ab" indicates an antigen where two different specificities are carried by the same antigen particle.
7. "A"–underlining increased concentration of reactant.
8. The immune systems aA and bB are serologically nonrelated.
9. Antigens "a" and "a_1" are serologically related.
10. Immune sera indicated "AA_1" obtained by immunizing with antigen "a."
11. Antibodies "A" are of higher specificity than antibodies "A_1."

11. The petri dishes are covered and from duplicate sets of dishes one is placed for two to three days at room temperature in a humid atmosphere and the other at 4°C for seven days.
12. Rinse the surface of the agar-covered petri dishes carefully with water to remove dust or other undissolved particles.
13. The immunoprecipitin patterns of the gel diffusion agar petri dishes are fixed and stained as described in procedure 87, Fixation and Staining of Immunoelectrophoresis Slides or Ouchterlony Gel Diffusion Patterns.

Evaluation

Although a large number of variations of the basic patterns of immunoprecipitin lines are possible, the simplest patterns are shown in Fig. 2.

Reagents

A. 1. 2.67 g $Na_2HPO_4 \cdot 2H_2O$ is dissolved up to 100 ml with H_2O.
   2. 2.04 g of KH_2PO_4 is dissolved up to 100 ml with H_2O. Add 67 ml of solution 1 to 33 ml of solution 2. This is 0.15 M phosphate buffer with a pH 7.1.

References

1. Ouchterlony, O., *Progress in Allergy* 5, 1 (1958), and 6, 30 (1962), S. Karger, Basel and New York.
2. Kabat, E. A. and M. M. Mayer, *Experimental Immunochemistry*, 2d ed., p. 78, Charles C. Thomas, Springfield, 1961.
3. *Lab Synopsis Diagnostic Reagents Bulletin*, edited by E. Haaf, p. 3, Behring Diagnostics, Irrc., 400 Crossways Park Drive, Woodbury, N.Y. 11797, 1969.

95 PASSIVE HEMAGGLUTINATION OF LIPOPOLYSACCHARIDE OR OTHER MATERIAL

Objective

After fixation of polysaccharide or similar antigens on the surface of red blood cells and the addition of homologous antisera, there occurs an agglutination of the red blood cells. Because the red blood cells do not participate in this reaction directly but are merely a marker of the reaction it is termed "passive." The advantage of this reaction is that it permits detection of the reaction of a nonprecipitating antibody with a hapten by visualization of the red blood cell agglutination. This method is very sensitive in the detection of small amounts of hapten or antigen.

Method

A. Preparation of the lipopolysaccharide.
  1. 6 mg lipopolysaccharide or other material is dissolved in 1 ml 0.25 M NaOH.
  2. Place in a 56°C water bath for 60 minutes.
  3. Cool to room temperature.
  4. Centrifuge at 2000 rpm for 15 minutes at 4°C.
  5. Aspirate the supernatant and discard the pellet.
  6. Neutralize the supernatant with 1 N acetic acid.
  7. Dialyze the supernatant against several changes of distilled water overnight.
  8. Lyophilize.
B. Coating of the erythrocytes with lipopolysaccharide or other material.
  9. Wash 1 ml of a 5% suspension of sheep red blood cells twice with phosphate-buffered saline, pH 7.2.*
  10. Centrifuge to obtain a packed pellet of sheep red blood cells and discard the supernatant.
  11. Dissolve 250 μg of lipopolysaccharide (step 8) or 1 mg Capsular Polysaccharide Antigen Preparation, procedure 5, in 10 ml saline.
  12. Add 50 μl of the packed sheep red blood cells (step 10) to the lipopolysaccharide (step 11). *Shake.*
  13. Incubate at 37°C with occasional shaking for 30 minutes.
  14. Wash three times with phosphate-buffered saline to remove excess antigen.
  15. Suspend the coated sheep red blood cells in 10 ml of phosphate-buffered saline (1:200).

---

*Sheep red blood cells can be purchased in Alsever's Solution from Grand Island Biological Company, 3175 Staley Road, Grand Island, New York 14072.

C. Hemagglutination test.

16. Heat inactivate the antisera at 56°C for 30 minutes.
17. Add to each of 16 tubes 0.2 ml of phosphate-buffered saline.
18. Add 0.2 ml of antiserum to the first tube and serially dilute the antiserum.
19. Add 200 μl of antigen-coated sheep red blood cells (step 15) to each tube. *Shake vigorously.*
20. Incubate at 37°C for 30 minutes.
21. Centrifuge at 1000 rpm for 1 minute.
22. Read the agglutination grossly. The titer is the reciprocal of the highest dilution which gives a visible agglutination.

References

1. Keogh, E. V., E. A. North and M. F. Warburton, *Nature* **161**, 687 (1948).
2. Neter, E., O. Westphal, O. Lüderitz and E. A. Gorzynski, *Ann. N.Y. Acad. Sci.* **66**, 141 (1956).

96 POLYACRYLAMIDE DISC GEL ELECTROPHORESIS

Objective

This is an analytical method of separating molecular components based primarily on their charge.

Method

1. With a Pasteur pipette fill each capped tube with lower gel solution A to a height of approximately 40 mm (0.5 ml).
2. Layer the surface of the gel solution with approximately 0.1 ml of water to ensure formation of a smooth flat gel surface.
3. Allow lower gel solution to polymerize for approximately 30 minutes.
4. Decant the layering water and any unpolymerized gel solution.
5. Carefully rinse the surface of each gel column with a small quantity of upper gel solution B. Examine and discard any tubes where surface mixing occurs.

6. Layer approximately 0.15 ml of upper gel solution B to each tube. Carefully overlay with approximately 0.1 ml of water.

7. Polymerize the upper gel solution for 15 minutes with fluorescent light 3 inches behind the gels.

8. Decant the layering water and any unpolymerized gel solution.

9. Remove the caps from the ends of the tubes.

10. Position tubes in electrophoresis chamber.

11. Fill upper and lower chambers of the electrophoresis apparatus with appropriate buffers.

12. In order to have proper layering and prevent mixing, the density of the sample is increased with sucrose to 5 to 10%. Carefully layer 50 μl of sample on top of the upper gel.

13. To mark the buffer front add 1.0 ml of bromophenol blue reagent E for the anionic system or methyl green reagent F for the cationic system to the upper buffer compartment. Mix.

14. Connect the power supply.* A current of 1.25 ma per column is applied until the marker has entered the stacking gel after which the current is increased to 2.5 ma per column for the remainder of the run. The front is marked by the dye marker and followed visually. The run is terminated when the front reaches a distance of about 2 mm from the bottom of the gel column (approximately 50 to 60 minutes).

15. After the electrophoresis is completed the power is turned off and the gels are immediately removed from the glass tubes by gently rimming with a 22 gauge needle. Water is forced through the needle during the rimming process to facilitate removal of the gel.

16. The gel is cut at the buffer front marked by the dye marker so that R_f (ratio of the distance migrated of a protein to the buffer front) measurements can be made.

17. Fix and stain the gels for 60 minutes in Coomassie Blue solution G or Amido Schwarz solution H. Rinse with water.

18. Destain by placing the gels in destaining solution I with several changes of destaining solution.†

19. The gels can be stored in parafilm sealed test tubes in 7.5% glacial acetic acid.

*The positive lead should be connected to the lower electrode and the negative lead to the upper electrode for the anionic system.

†The gels can also be destained electrophoretically using destaining solution I in the upper and lower buffer compartments for 2 hours at 5 ma per tube.

Reagents

Anionic Gel System–pH 9.3

A. Lower gel (Store all gel reagents in brown bottles in the refrigerator.)
   1. 30 g acrylamide
      0.8 g bisacrylamide
      Dilute to 100 ml with water.
   2. 18.15 g tris (hydroxymethyl) aminomethane
      24 ml 1 N HCl
      0.24 ml TEMED (or 0.4 ml at 0°C)
      Dilute to 100 ml with water (pH 9.1).
   3. 0.14 g ammonium persulfate (0.4 g at 0°C)
      Dilute to 100 ml with water (prepare fresh weekly).

   Add 10 ml (1) + 10 ml (2) + 20 ml (3) prior to use.

B. Upper gel (Store all gel reagents in brown bottles in the refrigerator.)
   1. 10 g acrylamide.
      0.8 g bisacrylamide.
      Dilute to 100 ml with water.
   2. 2.23 g tris (hydroxymethyl) aminomethane
      12.8 ml 1 M H_3PO_4
      0.1 ml TEMED
      Dilute to 100 ml with water (pH 6.7).
   3. 2 mg riboflavin
      Dilute to 100 ml with water.
   4. 80 mg ammonium persulfate.
      Dilute to 100 ml with water. (Prepare fresh weekly.)

   Add 10 ml (1) + 10 ml (2) + 10 ml (3) + 10 ml (4) prior to use.

C. Upper buffer
   5.16 g tris (hydroxymethyl) aminomethane
   3.48 g glycine
   Dilute to 1 liter with water (pH 8.91 at 25°C or at 0°C pH 9.64).

D. Lower Buffer
   14.5 g tris (hydroxymethyl) aminomethane
   60 ml 1 N HCl
   Dilute to 1 liter with water (pH 8.07 at 25°C or at 0°C pH 8.84).

Cationic Gel System–pH 4.3

A. Lower gel (Store all gel reagents in brown bottles in the refrigerator.)
1. 30 g acrylamide
   0.8 g bisacrylamide
   Dilute to 100 ml with water.
2. 24 ml 1 N KOH
   8.6 ml glacial acetic acid
   0.24 ml TEMED
   Dilute to 100 ml with water (pH 4.3).
3. 60 mg potassium persulfate
   2 mg riboflavin
   Dilute to 100 ml with water.

   Add 10 ml (1) + 10 ml (2) + 20 ml (3) prior to use.

B. Upper gel (Store all gel reagents in brown bottles in the refrigerator.)
1. 10 g acrylamide
   0.8 g bisacrylamide
   Dilute to 100 ml with water.
2. 48 ml 1 N KOH
   2.87 ml glacial acetic acid
   0.1 ml TEMED
   Dilute to 100 ml with water (pH 6.7).
3. 60 mg potassium persulfate
   2 mg riboflavin
   Dilute to 100 ml with water.

   Add 10 ml (1) + 10 ml (2) + 20 ml (3) prior to use.

C. Upper buffer
   8 ml glacial acetic acid
   31.2 g beta-alanine
   Dilute to 1 liter with water (pH 4.5).
   Dilute 1 : 10 for use.
D. Lower buffer
   43 ml glacial acetic acid
   120 ml 1 N KOH
   Dilute to 1 liter with water (pH 4.3).
   Dilute 1 : 10 for use.

E. Anionic System Bromophenol Blue Marker
5 mg bromophenol blue dissolved up to 500 ml with water.
F. Cationic System Methyl Green Marker
5 mg methyl green dissolved up to 500 ml with water.
G. Coomassie Blue Stain
1.25 g Coomassie brilliant blue
454 ml 50% methanol
46 ml glacial acetic acid. *Mix.*
Filter through Whatman No. 1.
H. Amido Schwarz Stain
7 ml glacial acetic acid
93 ml water
0.5 g Amido Schwarz. *Mix.*
Filter through Whatman No. 1.
I. Destaining solution
75 ml glacial acetic acid
50 ml methanol
875 ml water

References

1. Büchler Instruments, Inc., Fort Lee, New Jersey 07024.
2. Parish, C. R. and J. J. Marchalonis, *Anal. Biochem.* **34**, 436 (1970).

97 POLYACRYLAMIDE DISC GEL ELECTROPHORESIS WITH SODIUM DODECYL SULFATE

Objective

Oligomeric proteins in the presence of a detergent such as sodium dodecyl sulfate are split into polypeptide chains and can be separated using polyacrylamide disc gel electrophoresis. The polypeptide chains can be visualized by staining the gels and mobility of the moiety can be translated into its molecular weight using appropriate standards.

Method

1. Incubate a protein solution (200 to 600 μg/ml) for 2 hours at 37°C in buffer reagent B-1.

2. Dialyze the protein solution for several hours at room temperature against 500 ml of buffer reagent B-2.

3. For a run of 12 gels, 15 ml of gel buffer reagent C are deaerated and mixed with 13.5 ml of acrylamide gel reagent D. Continue deaeration.

4. Add 1.5 ml of ammonium persulfate reagent E.

5. Add 45 μl TEMED (N,N,N′,N′-tetramethylethylenediamine). *Mix*.

6. Fill each tube* with 2 ml of the solution.

7. Before the gel hardens, layer a few drops of water on top of the gel solution.

8. After 10 to 20 minutes an interface is visible indicating that the gels have solidified.

9. Just before use, the water layer is aspirated.

10. Place the tubes in the electrophoresis apparatus.

11. For each gel 3 μl of tracking dye reagent F, 1 drop of glycerol, 5 μl mercaptoethanol and 50 μl of buffer reagent B-2 are added to a small test tube. *Shake*.

12. Add 10 to 50 μl of the protein solution to the small tube in step 11. *Shake*.

13. Apply the protein solution (step 12) onto the top of the gel.

14. Fill the 2 compartments of the electrophoresis apparatus with gel buffer reagent C which has been diluted with an equal volume of water.

15. Perform electrophoresis at a constant current of 8 ma per gel with the positive electrode in the lower chamber.

16. It requires approximately 4 hours for the bromophenol blue dye marker to travel 3/4 of the distance through the gel and the power is then turned off.

17. The gels are removed from the tubes by squirting water from a syringe between the gel and glass wall.

18. The length of the gel and the distance the bromophenol blue dye moved through the gel are measured.

19. The gels are then placed in small test tubes filled with staining solution G for 2 to 10 hours.

20. Remove the gels from the staining solution. Rinse with water.

21. Place gels in destaining solution H for at least 30 minutes.
22. The gels are then further destained electrophoretically for 2 hours in the gel electrophoresis apparatus using destaining solution H.
23. Measure the length of the gel and the position of the blue protein zones.
24. Store the gels in acetic acid solution I.

Reagents

A. 1. 1.42 g Na_2HPO_4 diluted up to 1 liter with water (0.01 M).
   2. 1.38 g $NaH_2PO_4 \cdot H_2O$ diluted up to 1 liter with water (0.01 M).
   3. Add appropriate volumes 1 and 2 together until a pH of 7 is achieved.
B-1. Add 1 g of sodium dodecyl sulfate, and 1 ml of β-mercaptoethanol and dilute to 100 ml with reagent A.
B-2. Add 500 mg sodium dodecyl sulfate and 500 µl of β-mercaptoethanol and dilute to 500 ml with reagent A.
C. Gel buffer
   7.8 g $NaH_2PO_4 \cdot H_2O$
   38.6 g $Na_2HPO_4 \cdot 7H_2O$
   2 g sodium dodecyl sulfate
   Dilute up to 1 liter with water.
D. Acrylamide gel (final 10% solution)
   22.2 g acrylamide
   0.6 g methylenebisacrylamide
   Dilute up to 100 ml with water. Filter through Whatman No. 1 and store in a dark bottle at 4°C.
E. Ammonium persulfate
   1.5 g ammonium persulfate
   Dilute up to 100 ml with water.
   Prepare fresh weekly.
F. Bromophenol blue tracking dye
   50 mg bromophenol blue
   Dilute up to 100 ml with water (0.05%).
G. Staining solution
   1. 1.25 g Coomassie brilliant blue dissolved in 454 ml of 50% methanol and 46 ml of glacial acetic acid.
   2. Filter through Whatman No. 1.

H. Destaining solution
   75 ml glacial acetic acid
   50 ml methanol
   875 ml water
I. 7.5 ml glacial acetic acid added to 92.5 ml water (7.5%).

Evaluation

1. The gels swell in the acidic solution used for staining and destaining. In order to calculate the mobility of a protein, the length of the gel before and after staining as well as the mobility of the protein and marker dye must be considered.
   Calculate:

$$\text{Mobility} = \frac{\text{distance of protein migration after destaining}}{\text{length of gel after destaining}}$$

$$\times \frac{\text{length of gel before staining}}{\text{distance of dye marker migration before staining}}$$

2. The mobilities are plotted against the known molecular weights of protein standards on semi-logarithmic paper.
3. The molecular weight of an unknown protein can be obtained from the above standard curve.

Note:
The glass gel tubes are 10 cm long with an inner diameter of 6 mm. Before use they are soaked in cleaning solution, rinsed with water, and dried in an oven.

Reference

1. Weber, K. and M. Osborn, *J. Biol. Chem.* **244**, 4406 (1969).

98   PREPARATION OF ANTISERA

Objective

This is a schedule that is useful for preparing hyperimmune antisera in rabbits.

Method

1. Use normal, healthy New Zealand white rabbits weighing 2 to 3 kg which do not have natural antibodies against the antigen injected (tested by procedure 105, Tube Agglutination).
2. Use as antigen an amount of material in sterile saline that originated from a culture containing 4×10^9 bacteria/ml (e.g., see procedures 99, 100 and 106, Preparation of H Antigens, Preparation of O Antigens, and Vi Antigen Extraction).
3. Clip the hair from the ear of the rabbit and rub the ear vigorously with a xylene-soaked cloth. Wash the ear with sterile saline.
4. Inject intravenously into Vena marginalis (marginal ear vein) of the rabbit's ear using the following schedule:

Day 0	0.5 ml
Day 4	1.0 ml
Day 8	1.5 ml
Day 12	2.0 ml
Day 16	2.5 ml
Day 20	3.0 ml
6 days rest	
Day 26	Exsanguinate by cardiac puncture.

5. Let the blood stand for 90 minutes at room temperature to coagulate.
6. Centrifuge at 2000 rpm for 30 minutes and remove the serum supernatant.
7. Estimate the titer of the antisera by procedure 105, Tube Agglutination.
8. Lyophilize the sera or store at $-70°C$.

Note:
Generally 30 ml of sera can be obtained from each rabbit.

References

1. Ruschmann, E. and O. Lüderitz, *Zbl. Bakt. I. Abt. Orig.* **216,** 185 (1971).
2. Sedlák, J., *Enterobacteriaceae*, p. 227, S.Z.N. Publishing Co., Prague, 1955.

99 PREPARATION OF H ANTIGENS FOR ANTISERA PRODUCTION

Objective

This is a method for preparing H (flagellar) antigens from the appropriate strains of Enterobacteriaceae for production of hyperimmune sera in animals.

Method

1. To the appropriate 6-hour culture (4×10^9 cells/ml)* in nutrient broth add formaldehyde to a final concentration of 0.5%.
2. Let stand overnight at $37°C$.
3. The material is now ready for injection into rabbits to produce antibody (see procedure 98, Preparation of Antisera).

Reference

1. Sedlák, J. *Enterobacteriaceae*, p. 226, S.Z.N. Publishing Co., Prague, 1955.

100 PREPARATION OF O ANTIGENS FOR ANTISERA PRODUCTION

Objective

This is a method for preparing bacterial O antigens primarily from Enterobacteriaceae for production of hyperimmune sera in animals.

Method

1. Use 10^9 bacteria/ml suspended in saline from an agar culture (about a 24 hour culture grown at $37°C$).
2. Place in a boiling water bath for 2.5 hours.

*To increase the motility of the strain, if necessary, pass the bacteria five or six times through 0.2% agar medium. Increase of motility is necessary if the flagella are not well-developed.

3. Centrifuge at 7000 rpm for 30 minutes.
4. Suspend the pellet in 95% ethanol.
5. Incubate at 37°C for 4 hours.
6. Centrifuge at 7000 rpm for 30 minutes.
7. Wash the pellet twice with acetone (7000 rpm, 15 minutes at 4°C).
8. Suspend the washed pellet in a small amount of acetone.
9. Place in a dish and allow acetone to evaporate at 37°C overnight (alternatively one can evaporate the acetone with a stream of nitrogen at room temperature).
10. Suspend the acetone powder to the original concentration (as in step 1) in saline.
11. The material is now ready for injection into rabbits to produce antibody (see procedure 98, Preparation of Antisera).

Reference

1. Sedlák, J. *Enterobacteriaceae*, p. 266, S.Z.N. Publishing Co., Prague, 1955.

101 PYROGENICITY

Objective

In this method is described the measurement of pyrogenicity of lipopolysaccharide or other biological material which may cause a febrile response. It describes determination of the minimum pyrogenic dose. It also describes the method of evaluation of a material used parenterally to determine whether it is contaminated with pyrogenic products based on the *U.S. Pharmacopeia* standard.

Method

1. Use healthy New Zealand rabbits or American Dutch rabbits weighing 2 to 3 kg. (The rabbits should be acclimatized for one week prior to use.)
2. Twice in the week prior to beginning the experiment on the test material, pyrogen-free saline (10 ml/kg) should be injected to

determine that the rabbits will not give a false response. Rectal temperatures are measured from 40 minutes prior to injecting the pyrogen-free saline and for 5 hours after the injection at 60 minute intervals. The normal body temperature of rabbits is 38.2 to 39.8°C and the response should not be higher than 0.4°C. Reject any rabbits with a response greater than 0.4°C.

3. On the day of the experiment rectal temperatures are measured from 40 minutes prior to injecting the material to be tested. Reject any rabbits that have a fluctuation greater than 1°C from each other.

4. Inject the material preheated to 37°C intravenously, slowly (but not more than 2 minutes), into an ear vein at a dose of 10 ml/kg in pyrogen-free saline.

5. Use 10 ml/kg pyrogen-free saline (preheated to 37°C) as a control.

6. Measure the response in 3 rabbits injected with the material to be tested at hourly intervals for 3 hours.

7. Compare the response in 3 rabbits injected with pyrogen-free saline.

Evaluation

A: Pyrogenicity

1. If no rabbit shows an individual rise in temperature of 0.6°C or more above its respective control temperature and if the sum of the 3 temperature rises does not exceed 1.4°C, the material is apyrogenic.

2. If 3 rabbits give a response of 0.6°C or greater, the material is pyrogenic.

3. If 1 or 2 rabbits gives a temperature rise of 0.6°C above the control or the sum of the 3 temperatures exceeds 1.4°C, the experiment is repeated with 5 additional rabbits.

4. If not more than 3 rabbits from the 8 rabbits have an enhancement of 0.6°C or more and the sum of the 8 temperature rises does not exceed 3.7°C, the preparation is not considered pyrogenic. However, if 4 or more of the 8 rabbits have an enhancement of 0.6°C the material is considered pyrogenic.

B: MPD-3 (Minimum Pyrogenic Dose at 3 hours)

1. For quantitation of pyrogenic activity of lipopolysaccharides, lipid A and similar material, the febrile response is generally

expressed as the MPD-3 (minimum pyrogenic dose in 3 hours). It is defined as the smallest amount of toxin giving a mean rise of 0.6°C, 3 hours after intravenous injection. Each material is tested at 3 to 6 concentrations (and each concentration is tested in 5 rabbits weighing 1.0 to 1.2 kg). The mean febrile response at 3 hours is plotted against the log of the concentration of the material. The best line is drawn to 0.6°C and the concentration of toxin at the intercept represents the MPD-3.

References

1. Watson, D. W. and Y. B. Kim, *J. Exp. Med.* **118**, 425 (1963).
2. *United States Pharmacopeia*, p. 886, edit. XVIII, 1970.

102 QUANTITATIVE MICROPRECIPITATION

Objective

In this reaction a soluble polyvalent antigen creates with a bivalent antibody, a nonsoluble precipitate. The precipitate is washed and the amount of antibody protein which is in the precipitate is determined. This permits an estimation of the concentration of antibody protein against a soluble antigen in the serum.

Method

1. Centrifuge the antiserum (see procedure 98, Preparation of Antisera) for 30 minutes at 5000 rpm.
2. 10 μl of the polyvalent polysaccharide antigen (1%) (see procedure 27, Purification of Lipopolysaccharide [modified Westphal]) or a protein antigen is added to the first of 10 tubes. The tubes should be 27 mm long and have an inner diameter of 4 mm. (This should be repeated in triplicate to insure accuracy.)
3. Prepare two-fold serial dilutions of the antigen (from step 2), i.e., add 10 μl saline to each tube and transfer 10 μl from the first tube to each successive tube.
4. Blanks of saline substituted for antigen should be included.
5. Add 10 μl of antiserum to each tube. *Shake.*
6. Incubate in a 37°C water bath for 60 minutes.
7. Let stand for seven days at 2°C. *Shake gently twice daily.*

8. Centrifuge at 3800 g for 60 minutes at 2°C and carefully decant the supernatant with a micropipet and discard it.
9. Wash the precipitate three times with 15 μl saline.
10. Dry the precipitate in a vacuum desiccator over $CaCl_2$.
11. Add 20 μl 1 N NaOH to each tube. *Shake.*
12. Let stand at room temperature for 30 minutes. *Shake.*
13. Add 100 μl water. *Shake.*
14. Determine the protein by the Lowry Method (procedure 73).

Evaluation

1. Plot the amount of antigen against the protein concentration in the precipitate. This is the precipitation curve.
2. The maximum of the precipitation curve is the equivalence point and the protein concentration at the equivalence point is the total antibody present if a lipopolysaccharide antigen (which contains no protein) is used.
3. However, if a protein antigen is used, the amount of antigen has to be deducted from total protein at the equivalence point.

References

1. Heidelberger, M. and F. E. Kendall, *J. Exp. Med.* **62**, 697 (1935).
2. Kabat, E. A. and M. M. Mayer, *Experimental Immunochemistry*, 2nd ed., p. 76, Charles C. Thomas, Springfield, 1961.

103   SIMPLE DIFFUSION (OUDIN)

Objective

This test reveals the presence of an antigen-antibody reaction in a narrow test tube containing agar, i.e., simple diffusion. With this procedure the antigen-antibody precipitation occurs in the agar gel medium. Due to the different diffusion rates of the heterogenous antigens, the differentiation of the various antigens is made possible. If antibody is in excess the band of precipitation rises but if antigen is in excess the band of precipitation falls. If the system is balanced the precipitation band remains at the place it originates.

Method

1. The test is carried out in capillary tubes 100 mm by 3 mm inner diameter which have been internally coated with a film of agar.
2. The tubes are coated with agar as follows:
   a. Suspend 1 g of agar in 100 ml water in a beaker and carefully heat to boiling to dissolve the agar.
   b. Place the capillary tubes in the agar solution.
   c. Cover the beaker and autoclave for 15 minutes at 15 lbs pressure (121°C).
   d. Allow the beaker to cool to 45 to 50°C and remove the excess agar from the tubes by shaking them.
   e. Place the agar-coated capillary tubes in a clean beaker in a hot air oven at 56°C overnight to dry.
3. Prepare serial two-fold dilutions of the antisera by adding 100 μl of buffer reagent B to each of 6 tubes and 100 μl of antisera to the first tube. *Shake.* Remove 100 μl from the first tube and serially dilute.
4. Immediately prior to addition to the capillary tubes, add 100 μl agar reagent A to each of the serially diluted antisera tubes. (Both antisera and agar should be about 45°C prior to mixing.)
5. Dispense the 200 μl of antisera agar mixture to the capillary tubes (which comes to a height of about 30 mm) with care to avoid air bubbles. This can be conveniently done with a hot Pasteur pipette.
6. Allow to solidify at room temperature and then place in a refrigerator.
7. Carefully layer onto the agar in the capillary tube 200 μl of a solution of antigen. (It may be convenient to also incorporate the antigen in agar reagent A.)
8. Seal the tubes with rubber caps or rubber stoppers and melted paraffin around the stoppers to prevent evaporation.
9. Incubate at room temperature for 24 to 72 hours (although some precipitates may require four to ten days to appear).
10. Visualization of the precipitate is facilitated by allowing a strong light to pass through the tubes obliquely.

Reagents

A. 0.60 g agar (purified), e.g., ionagar, 0.85 g NaCl, 100 ml water. Autoclave for 15 minutes at 15 lbs pressure (121°C). Add 20 mg merthiolate.

B. 0.06 g diethylbarbituric acid, 0.04 g sodium 5,5-diethylbarbiturate, 20 mg merthiolate, 100 ml water, pH 7.4.

Note:

Instead of using this isotonic buffer for the serial dilutions of the antisera, a normal sera of the same species (unimmunized) can be utilized. Thus the serial dilution of the antisera can be performed in normal unimmunized sera as described in step 3.

Comments

1. The antigen must be well dissolved generally in saline.
2. This system can be reversed by using serially diluted antigen in the agar and overlaying the agar with the antisera.
3. As controls, the antisera should be omitted from the agar and only buffer utilized. Also, antigen should be omitted and saline used.
4. The distance, h, of the front of the precipitate from the boundary surface of agar is proportional to the square root of the diffusion time, t:

$$h = k \times \sqrt{t}$$

Since the migration velocity of the antigen, $k = \dfrac{h}{\sqrt{t}}$, is proportional to the logarithm of the antigen concentration, the relationship between antigen concentration, C_{Ag}, and the distance h through which the precipitate migrates may be expressed as follows:

$$\log C_{Ag} = a \times \frac{h}{\sqrt{t}} + b$$

Therefore, if C_{Ag} (logarithmic) is plotted against $\dfrac{h}{\sqrt{t}}$ (linear) on semi-logarithmic paper a straight line is obtained. Using this standard curve unknown concentrations of antigen may be estimated.

Reference

1. Oudin, J., *Methods in Medical Research*, **5**, 335 (1952).

104 SLIDE AGGLUTINATION

Objective

This is a method for qualitative determination of a corpuscular antigen to an antiserum. It is often used as a diagnostic tool for the identification of an unknown corpuscular antigen.

Method

1. Prepare on a watchglass (Fig. 1) a suspension of antigen in saline. (If bacteria are utilized, use about 10^9 cells/ml.)

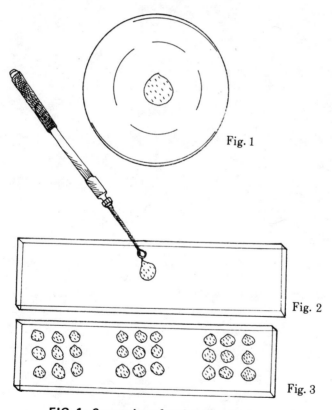

Fig. 1

Fig. 2

Fig. 3

FIG. 1 Suspension of antigen in saline.
FIG. 2 Preliminary agglutination test.
FIG. 3 Serial dilution for slide agglutination.

2. Do a preliminary test by mixing 1 loopful of antigen with 1 loopful of antisera on a slide (Fig. 2). Evaluate for agglutination.
3. If there is agglutination then serially dilute the antisera in the following manner (Fig. 3): Add patterns of 9 loopfuls of antigen on the slide. Then add 1 loopful of antisera to each pattern, mix thoroughly and then transfer 1 loopful to the next 9 spot pattern. Serially dilute the antigen until no further agglutination occurs.
4. The last dilution to have agglutination is the titer.

Evaluation

1. After transferring the loop for the serial dilutions, mix well with the loop as a stirrer. Tilt the slide back and forth gently about ten times. Agglutination must occur within 3 minutes for the test to be considered positive.
2. It is best to evaluate for agglutination by visually examining the slide against a black background.

Reference

1. Sedlák, J., *Enterobacteriaceae*, p. 57, S.Z.N. Publishing Co., Prague, 1955.

105 TUBE AGGLUTINATION

Objective

This is a method for estimation of the titer of specific agglutinating antibodies in serum. It is a quantitative method that is a great deal more sensitive than slide agglutination (procedure 104).

Method

O Agglutination (Somatic Antigen)

1. Dilute the antisera 1 : 10 with saline, i.e., 1 ml of antiserum to 9 ml saline.
2. To the first tube add 0.2 ml of the diluted antiserum and to each of 8 successive tubes add 0.1 ml of saline. Serially dilute 0.1 ml of

the antiserum to each of the 8 tubes. Shake well after each dilution. (i.e., 1:20, 1:40, 1:80, etc.)
3. Add to each tube 0.1 ml of antigen. (If bacteria are used they should be at a concentration of 10^9 cells/ml.)
4. Incubate in a 50°C water bath for 20 hours.

H Agglutination (Flagellar)

1. Dilute the H antiserum 1:50, i.e., 0.1 ml of antiserum, 4.9 ml saline.
2,3. Proceed as in steps 2 and 3 above (i.e., 1:100, 1:200, 1:400 etc.).
4. Let the tubes stand in a 50°C water bath for 120 minutes and then at room temperature for 20 hours.

Vi (K) Agglutination

1. Dilute the Vi antisera 1:5, i.e., 2 ml antiserum + 8 ml saline.
2,3. Proceed as in steps 2 and 3 above (i.e., 1:10, 1:20, 1:40 etc.).
4. Incubate in 37°C water bath for 120 minutes and at room temperature for 20 hours.

Controls

1. Positive Control
   A. Antigen control—Use a commercially available antigen with your antiserum.
   B. Antiserum control—Use a commercially available antiserum with your antigen.
   C. Combined control—Use a commercially available antigen and antiserum.
2. Negative Control
   A. Use 0.1 ml of saline and 0.1 ml antigen to determine that there is no nonspecific agglutination.

Evaluation

Shake the tubes individually by a quick flick of the wrist and allow about 10 seconds for settling and compare with the positive and negative controls.

1. O antigen–A seed-like sediment with a clear supernatant appears.

2. H antigen–A cloudy cotton-like sediment which disperses easily appears.
3. Vi antigen—A thin sediment with well-defined serrated edges and a clear supernatant which disperses easily appears.

Reference

1. Sedlák, J., *Enterobacteriaceae*, p. 57, S.Z.N. Publishing Co., Prague, 1955.

106 Vi ANTIGEN EXTRACTION (FOR HEMAGGLUTINATION)

Objective

This is a method for preparing Vi antigen (or K antigen of enteritic *E. coli* strains). The principle of the method is the adsorption of the Vi extract on a nonspecific carrier, specifically human erythrocytes of the O blood group. The sensitized blood cells with the antigen extract are agglutinogens for analyzing specific antibodies.

Method

1. Use a high layer of 1.5% nutrient broth agar for cultivation of the bacteria. Grow for 20 hours at 37°C.
2. Remove the cells from the plates with 75% ethanol.
3. Let stand at room temperature for 30 minutes.
4. Centrifuge for 45 minutes at 3000 rpm at 4°C.
5. Dry the pellet in a dish for 4 hours at 37°C.
6. Suspend the cells in a two-fold volume (compared to the weight of the pellet) with water.
7. Incubate at 37°C for 4 hours.
8. Let stand at 4°C for 16 hours.
9. Centrifuge for 60 minutes at 12,000 rpm at 4°C.
10. The supernatant contains the Vi extract for sensitization of the erythrocytes to be used for hemagglutination (see procedure 95, Passive Hemagglutination of Lipopolysaccharide or Other Material).

Note:

The Vi antigen is present in *Salmonella paratyphi C, Salmonella typhi,* and some strains of *Citrobacter.* Similarly, one can prepare capsular antigens from *E. coli* B_4, B_5, and B_6.

Reference

1. Sedlák, J., *Enterobacteriaceae*, p. 227, S.Z.N. Publishing Co., Prague, 1955.

Author Index

Note: Boldface indicates procedure number.

157

Subject Index

Note: Boldface indicates procedure number.